VCE Units 3 & 4
PSYCHOLOGY

KRISTY KENDALL

2023–2027 STUDY DESIGN • 2023–2027 STUDY DESIGN •

A+

+ 10 topic tests
+ two complete practice exams
+ detailed, annotated answers

PRACTICE
EXAMS

A+ VCE Units 3 & 4 Psychology Practice Exams
3rd Edition
Kristy Kendall
ISBN 9780170465205

Publisher: Alice Wilson
Series editor: Catherine Greenwood
Copyeditor: Natalie Craig
Contributing author (multiple-choice question answers): Gabrielle Painter
Consultant: Meredith McKague
Reviewers: Matt Brinson, Kirsten Horne
Series text design: Nikita Bansal
Series cover design: Nikita Bansal
Series designer: Cengage Creative Studio
Artwork: MPS Limited
Production controller: Karen Young
Typeset by: Nikki M Group Pty Ltd

Any URLs contained in this publication were checked for currency during the production process. Note, however, that the publisher cannot vouch for the ongoing currency of URLs.

Acknowledgements

Selected VCE examination questions and extracts from the VCE Study Designs are copyright Victorian Curriculum and Assessment Authority (VCAA), reproduced by permission. VCE ® is a registered trademark of the VCAA. The VCAA does not endorse this product and makes no warranties regarding the correctness or accuracy of this study resource. To the extent permitted by law, the VCAA excludes all liability for any loss or damage suffered or incurred as a result of accessing, using or relying on the content. Current VCE Study Designs, past VCE exams and related content can be accessed directly at www.vcaa.vic.edu.au.

© 2023 Cengage Learning Australia Pty Limited

For product information and technology assistance,
in Australia call **1300 790 853**;
in New Zealand call **0800 449 725**

For permission to use material from this text or product, please email
aust.permissions@cengage.com

ISBN 978 0 17 046520 5

Cengage Learning Australia
Level 7, 80 Dorcas Street
South Melbourne, Victoria Australia 3205

Cengage Learning New Zealand
Unit 4B Rosedale Office Park
331 Rosedale Road, Albany, North Shore 0632, NZ

For learning solutions, visit **cengage.com.au**

Printed in China
2 3 4 5 6 7 26 25 24 23

CONTENTS

HOW TO USE THIS BOOK...........................iv
A+ DIGITAL FLASHCARDS..........................v
ABOUT THE AUTHOR..............................vi

HOW DOES EXPERIENCE AFFECT BEHAVIOUR AND MENTAL PROCESSES?

Area of Study 1

Test 1: Nervous system functioning 2
Test 2: Stress as an example of a psychobiological process 8

Area of Study 2

Test 3: Models to explain learning 15
Test 4: The psychobiological process of memory 21

HOW IS MENTAL WELLBEING SUPPORTED AND MAINTAINED?

Area of Study 1

Test 5: The demand for sleep 28
Test 6: Importance of sleep to mental wellbeing 34

Area of Study 2

Test 7: Defining mental wellbeing 41
Test 8: Application of a biopsychosocial approach to explain specific phobia 48
Test 9: Maintenance of mental wellbeing 55

Area of Study 3

Test 10: Experimental design 62

PRACTICE EXAMS

Practice Exam 1 71
Practice Exam 2 93

ANSWERS....................................115

HOW TO USE THIS BOOK

The *A+ VCE Psychology* resources are designed to be used year-round to prepare you for your VCE Psychology exam. *A+ VCE Psychology Units 3 & 4 Practice Exams* includes 10 topic tests and two practice exams, plus detailed answers for all questions in this resource. This section gives you a brief overview of the features included in this resource.

Topic tests

Each topic test addresses one key knowledge area of Units 3 and 4 of the VCE Psychology Study Design. The order of tests follows the same sequence as in the syllabus.

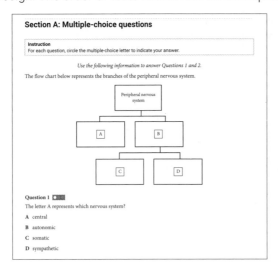

Practice exams

Both practice exams cover all key knowledge of the VCE Psychology Study Design. The practice exams have perforated pages so that you can remove them from the book and practise under exam-style conditions.

Answers

Answers to topic tests and practice exams are supplied at the back of the book. They have been written to reflect a high-scoring response and include explanations of what makes an effective answer.

Explanations

The answers section includes explanations of each multiple-choice option, both correct and incorrect, and explanations to written response items, which explain what a high-scoring response looks like and signposts potential mistakes.

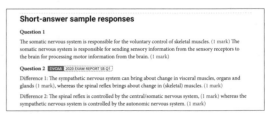

Icons

You will notice the following icons in the tests and practice exams.

©VCAA 2020 SA Q4

This icon appears with official past VCAA exam questions.

These icons indicate whether the question is easy, medium or hard.

Aboriginal peoples and Torres Strait Islander peoples

Australia is home to two distinct, broad groups of First Nations peoples. These are Aboriginal peoples and Torres Strait Islander peoples. Collectively, the phrase 'Aboriginal and Torres Strait Islander peoples' refers to the group of nations, cultures and languages across Australia and throughout the Torres Strait. We speak of 'peoples' (plural) to recognise that there are many different nations, cultures and language groups, not just one Aboriginal and Torres Strait Islander culture or identity.

The term 'Aboriginal' refers broadly to the nations and custodians of mainland Australia and most of the islands, including Tasmania, Fraser Island, Palm Island, Mornington Island, Groote Eylandt, Bathurst Island and Melville Island.

The term 'Torres Strait Islander' refers broadly to the peoples of at least 274 small islands between the northern tip of Cape York in Queensland and the south-west coast of Papua New Guinea.

Visit the Australian Institute for Aboriginal and Torres Strait Islander Studies (AIATSIS) website to view the map of Indigenous Australia, which shows the major language groups and their rough geographical boundaries.

A+ VCE Psychology Study Notes

A+ VCE Psychology Practice Exams can be used independently or alongside the accompanying resource, *A+ VCE Psychology Study Notes*. The Study Notes include topic summaries and exam practice for all key knowledge in the VCE Psychology syllabus that you will be assessed on during the exam, as well as detailed revision and exam preparation advice to help you get ready for the exam.

A+ DIGITAL FLASHCARDS

Revise key terms and concepts online with the A+ Flashcards.
Just scan the QR code or type the URL into your browser to access them.
Note: You will need to create a free NelsonNet account.

https://get.ga/
aplus-vce-psych-u34

ABOUT THE AUTHOR

Kristy Kendall

Kristy Kendall graduated from Monash University with a Bachelor of Arts, majoring in Psychology. She has also completed her Master of Education, focusing on the development of executive functions in the adolescent brain. She is currently the Principal of Toorak College in Mt Eliza and is the key lecturer in Psychology on the online resource Edrolo.

Kristy is a former VCAA examiner and has published works in VCE Psychology, including *A+ VCE Psychology Practice Exams*, *Nelson's VCE Psychology Research Methods Workbook* and many other VCE Psychology titles.

9780170465205

UNIT 3
HOW DOES EXPERIENCE AFFECT BEHAVIOUR AND MENTAL PROCESSES?

Area of Study 1:

Test 1	Nervous system functioning	2
Test 2	Stress as an example of a psychobiological process	8

Area of Study 2:

Test 3	Models to explain learning	15
Test 4	The psychobiological process of memory	21

AREA OF STUDY 1
TEST 1: NERVOUS SYSTEM FUNCTIONING

Section A: 20 marks. Section B: 20 marks. Total marks: 40.
Suggested time: 50 minutes

Section A: Multiple-choice questions

Instruction
For each question, circle the multiple-choice letter to indicate your answer.

Use the following information to answer Questions 1 and 2.

The flow chart below represents the branches of the peripheral nervous system.

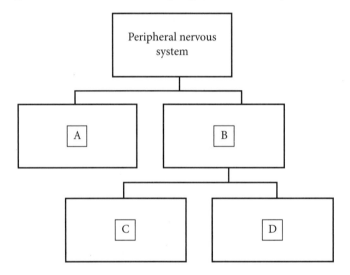

Question 1

The letter A represents which nervous system?

A central

B autonomic

C somatic

D sympathetic

Question 2

The letter B represents which nervous system?

A central

B autonomic

C somatic

D sympathetic

Question 3

Which of the following statements best describes a feature of the autonomic nervous system?

A The autonomic nervous system involves voluntary control of skeletal muscles.

B The autonomic nervous system has two subdivisions: the somatic and parasympathetic nervous systems.

C The autonomic nervous system integrates sensory and motor information.

D The autonomic nervous system is responsible for regulating physiological responses that do not require conscious control.

Question 4

Which combination of events places the body in a heightened state of arousal?

A increased heart rate, decreased lung capacity

B decreased saliva, dilated pupils

C increased lung capacity, increased saliva

D constricted pupils, increased heart rate

Question 5

Following heightened arousal, the parasympathetic nervous system calms the body. Which of the following changes occurs?

A heart rate increases

B digestion increases

C release of bile is inhibited

D production of saliva decreases

Question 6

Which of the following responses does not involve the brain?

A breathing

B reading

C talking

D spinal reflex

Question 7 ©VCAA 2019 SA Q2

When someone pricks their finger and immediately withdraws it, their response demonstrates

A the adaptive nature of the human nervous system.

B how the spinal cord makes decisions about movement.

C the conscious response involved in the coordination of the reflex.

D the role of the brain in the responses of the autonomic nervous system.

Question 8

Neural communication occurs through the use of

A electrical energy only.

B electrochemical energy.

C chemical energy only.

D neural energy.

Question 9

A synapse is integral to neural communication. What is it?

A a gap between two neurons

B a type of neurotransmitter

C an action potential

D the resting potential

Question 10

Which of the following is most typically the role of an excitatory neurotransmitter?

A to stabilise mood

B to inhibit a response

C to calm someone down

D to stimulate a response

Question 11

The changing of neurons in the brain as a result of new experiences is known as

A hippocampus manipulation.

B neuroplasticity.

C synaptic formation.

D neurotransmission.

Question 12

According to research, in which of the following areas of the neuron does memory formation occur?

A dendrite

B synapse

C axon

D myelin sheath

Question 13

During learning, neural pathways are strengthened when

A the presynaptic neuron is firing, but the postsynaptic neuron is not.

B the postsynaptic neuron is firing, but the presynaptic neuron is not.

C the presynaptic neuron and postsynaptic neuron are firing at the same time.

D neither the presynaptic neuron nor the postsynaptic neuron is firing.

Question 14

Which of the following statements about learning is true?

A There is a change in neurons when learning takes place.

B Neurons are not affected by learning.

C Neurons only change temporarily when learning takes place.

D The number of neurons decreases when learning takes place.

Question 15 ⬤⬤⬤

The role of the neuron is to

A travel towards the brain.

B travel away from the brain.

C transmit sensory messages to the brain.

D transmit motor messages to the brain.

Question 16 ⬤⬤⬤

The body has many protective mechanisms to help optimise its chance for survival. For example, a person who touches an extremely hot or cold object can have an unconscious response, withdrawing their hand before pain has even been registered by their brain. This is probably the result of the

A peripheral nervous system.

B autonomic nervous system.

C central nervous system.

D sympathetic nervous system.

Question 17 ⬤◐◐

Gamma-aminobutyric acid (GABA) is a

A neurotransmitter.

B hormone.

C neuron.

D part of the brain.

Question 18 ⬤⬤◐

The release of gamma-aminobutyric acid (GABA) helps to

A create a sensation of pain.

B decrease a feeling of anxiety.

C remove a sensation of pain.

D excite neural impulses.

Question 19 ⬤◐◐

Glutamate is an excitatory neurotransmitter essential for

A learning and memory.

B motor movement.

C long-term depression.

D reward.

Question 20 ⬤⬤◐

Which of the following effects are not commonly associated with dopamine?

A memory

B reward

C pleasure

D motivation

Section B: Short-answer questions

Instruction
Answer all questions in the spaces provided.

Question 1 (2 marks) ●●

Describe two functions of the somatic nervous system.

Question 2 (4 marks) ©VCAA 2020 SB Q1 ●●

Outline two differences between the sympathetic nervous system response and the spinal reflex.

Difference 1 _____ 2 marks

Difference 2 _____ 2 marks

Question 3 (3 marks) ●●●

Neural communication is essential for coordinating every thought and action. Write a detailed description of how two neurons communicate by transmitting and receiving information.

Question 4 (3 marks)

a What is long-term potentiation (LTP)?   1 mark

_____

_____

b Discuss two physiological changes that occur in neurons when LTP takes place.   2 marks

_____

_____

_____

Question 5 (5 marks)

a Liam is 3 years old and a happy and healthy child. His brain is developing at a rapid rate.
Explain the role of developmental plasticity in this.   2 marks

_____

_____

_____

_____

b If Liam suffered a brain injury later in life, he would rely on adaptive plasticity. What is
adaptive plasticity?   2 marks

_____

_____

_____

c Identify the optimal time for adaptive plasticity across the life span.   1 mark

_____

_____

_____

Question 6 (3 marks)

Gambling is a multi-billion-dollar industry in Australia. This is partly due to the large number
of regular gamblers for whom gambling has become an addiction. Describe a biological theory
for why gambling can become an addiction.

_____

_____

_____

_____

AREA OF STUDY 1
TEST 2: STRESS AS AN EXAMPLE OF A PSYCHOBIOLOGICAL PROCESS

Section A: 20 marks. Section B: 20 marks. Total marks: 40.
Suggested time: 50 minutes

Section A: Multiple-choice questions

Instruction
For each question, circle the multiple-choice letter to indicate your answer.

Question 1

What is the name of the substance that is secreted to regulate changes in the body due to stress?

A GABA

B glutamate

C dopamine

D cortisol

Question 2

Stress that produces high levels of arousal over time is commonly known as

A eustress.

B chronic stress.

C acute stress.

D a stressor.

Question 3

Just before he races, Indra gets butterflies in his stomach. He reports that this feeling is a sign that he is 'in the zone' and ready to go. This is most likely an example of

A an anxiety disorder.

B distress.

C chronic stress.

D eustress.

Question 4

According to the theory relating to the fight–flight–freeze response, when is someone most likely to freeze?

A when they are tired

B when they are injured

C when they are stressed

D when they can't choose between fight or flight

Question 5

The fight–flight–freeze response is a response to what kind of stress?

A   acute

B   chronic

C   eustress

D   emergency

Question 6

Miss Bell is always working hard. During the year she has to mark 300 SACs, coach two tennis teams and run all the social activities for the Year 12 students. Miss Bell is able to cope for a while, but eventually succumbs to illness, resulting in a flu she cannot fight off. When her breathing becomes problematic and she is eventually admitted to hospital, Miss Bell is most probably in the _____ stage of _____.

A   resistance; the General Adaptation Syndrome

B   resistance; the fight–flight–freeze response

C   exhaustion stage; the General Adaptation Syndrome

D   exhaustion stage; the fight–flight–freeze response

Question 7

In the alarm reaction stage of the General Adaptation Syndrome, the body's resistance to stress

A   rises above normal level, then drops below.

B   rises above normal level.

C   drops below normal level.

D   drops below normal level, then rises above.

Question 8

Which of the following characteristics is **not** typical of the exhaustion stage of the General Adaptation Syndrome?

A   increased immunity

B   colds and flu

C   stomach ulcers

D   high blood pressure

Question 9   ©VCAA 2019 SA Q9

Jamie is experiencing a constant state of stress and has also caught a cold.

Which of the following most accurately identifies the stage of Selye's General Adaptation Syndrome that Jamie is in and the reason that supports this stage?

	Stage	Reason
A	shock	Jamie's immune system is immobilised so his body can fight the stressor.
B	resistance	Continued cortisol release weakens Jamie's immune system, resulting in his body being unable to fight the cold.
C	exhaustion	Jamie's body's resources are depleted, resulting in vulnerability to a range of serious physical disorders.
D	resistance	Increased adrenaline in Jamie's bloodstream results in his body becoming susceptible to illnesses.

Use the following information to answer Questions 10–12.

Masako was anxious about and excited to be competing in the last baseball game before the finals. If her team won, it would progress to the finals. Masako was new to the sport and doubted her abilities but had practised a lot and carefully listened to her coach's tips. She had also decided that this game would help increase her skills. When it came time for Masako to bat, she was concentrating so closely on the ball that she blocked out the crowd cheering her on.

Question 10 ©VCAA 2019 SA Q4 ●●

Masako was most likely experiencing eustress because

A she was doubting her abilities.

B she felt excited about progressing to the finals if her team won.

C the stress of doing a good job interfered with her concentration.

D she felt nervous about not having much experience playing baseball.

Question 11 ©VCAA 2019 SA Q5 ●●

Which of the following identifies the functioning of Masako's autonomic nervous system and a resulting physiological response when she was preparing to bat?

	Autonomic nervous system functioning		Physiological response
	Parasympathetic nervous system	**Sympathetic nervous system**	
A	active	inactive	decreased salivation
B	non-dominant	dominant	increased blood pressure
C	inactive	dominant	movement of skeletal muscles
D	inactive	active	constricted pupils

Question 12 ©VCAA 2019 SA Q6 ●●●

According to Lazarus and Folkman's Transactional Model of Stress and Coping, an example of Masako undertaking primary appraisal would be if she thought

A of the crowd cheering her on.

B of the tips given to her by her coach.

C of the situation as good practice for the finals.

D that she had practised enough to hit the ball a long way.

Question 13 ●●

In Lazarus and Folkman's Transactional Model of Stress and Coping, the first stage is the primary appraisal stage, when a person decides

A whether they have appropriate resources to cope with the situation.

B who they should turn to for help.

C whether it is a stressful situation or not.

D whether or not the demands are too great to cope with.

Question 14

According to Lazarus and Folkman's Transactional Model of Stress and Coping, at which stage does a person evaluate available resources to help determine their level of stress?

A during primary appraisal

B during secondary appraisal

C during resistance

D during exhaustion

Question 15

If someone assesses a situation as stressful, which of the following is perceived as the most positive form of stress?

A harm

B loss

C threat

D challenge

Question 16

According to Lazarus and Folkman's Transactional Model of Stress and Coping, stress is greatest when

A there are too many demands.

B there are no demands.

C there are too many demands for available resources.

D there are too few demands for available resources.

Question 17

Which of the following relationships explains the impact of microbiota on psychological health?

A HPA axis

B gut–brain axis

C spinal reflex

D General Adaptation Syndrome

Question 18

What are microbiota?

A living micro-organisms

B neurons

C neurotransmitters

D stress cells

Question 19 ☐■■

There are many different means of coping with stressful situations. One strategy involves thinking about the cause of the stress and implementing a plan to resolve it. For example, someone who needs more money may decide to sell some belongings on eBay. This coping mechanism is known as

A emotional forecasting.

B problem-focused coping.

C emotion-focused coping.

D eustress.

Question 20 ■■■

Which of the following best describes why exercise is a healthy strategy for helping people deal with stress?

A It leads to a decrease in endorphins.

B It utilises unused adrenaline.

C It increases cortisol.

D It leads to exhaustion.

Section B: Short-answer questions

> **Instruction**
> Answer all questions in the spaces provided.

Question 1 (5 marks) ■■■

Name two physiological and two psychological stress reactions, then clearly outline the difference between these two types of reactions.

_____

_____

_____

_____

_____

Question 2 (3 marks) ■■■

a List one physiological response that increases when the fight–flight–freeze response is activated, and one physiological response that decreases.   2 marks

Increases _____

_____

Decreases _____

_____

b Explain why some responses increase and others decrease when the body is in a state of heightened arousal.   1 mark

_____

_____

_____

Question 3 (5 marks)

a Ruby is studying Fine Art at university but is finding the course difficult. She has many friends in her classes and is happy once she gets to university, but getting there can be difficult because of heavy traffic, and she feels isolated when she studies at home. Use Lazarus and Folkman's Transactional Model of Stress and Coping to explain how Ruby may evaluate the course in both the primary and secondary appraisal stages.   2 marks

Primary appraisal _____

_____

_____

_____

Secondary appraisal _____

_____

_____

_____

b Explain how primary and secondary appraisals of a situation may differ for two different people, according to Lazarus and Folkman's Transactional Model of Stress and Coping.   3 marks

_____

_____

_____

_____

_____

_____

Question 4 (3 marks) ⬤⬤⬤

Explain how the gut–brain axis can have a positive or negative effect on a person's mental health and wellbeing, and why the gut is often referred to as the 'second brain'.

_____

_____

_____

_____

_____

_____

Question 5 (4 marks) ⬤⬤⬤

Annabelle owns a small café by the beach and has felt overwhelmed at work recently. She is managing staff shortages, new government regulations and a fall in customers. Discuss how she may use approach or avoidance strategies for coping. Use practical examples and explain the relative effectiveness of each strategy.

_____

_____

_____

_____

_____

_____

_____

AREA OF STUDY 2
TEST 3: MODELS TO EXPLAIN LEARNING

Section A: 20 marks. Section B: 20 marks. Total marks: 40.
Suggested time: 50 minutes

Section A: Multiple-choice questions

Instruction
For each question, circle the multiple-choice letter to indicate your answer.

Question 1

Learning is defined as

A an active information-processing system.

B a brief change in behaviour due to interaction with the environment.

C a relatively permanent change in behaviour that occurs as a result of experience.

D a change in behaviour that always occurs intentionally.

Question 2

During the conditioning phase of classical conditioning, there is repeated association between the

A unconditioned stimulus and unconditioned response.

B unconditioned stimulus and conditioned response.

C unconditioned stimulus and neutral stimulus.

D unconditioned response and neutral stimulus.

Use the following information to answer Questions 3 and 4.

When Kim was young, she was playing with some yellow balloons at her sister's birthday party. Another child came and popped all the balloons. The loud noise scared Kim and she cried. After the incident, Kim would cry whenever she saw yellow balloons. Years later, as a result, Kim was unable to accept a job at a children's play centre because of her fear of yellow balloons.

Question 3

In this example, the conditioned stimulus was

A yellow balloons.

B crying from the loud noise.

C crying at yellow balloons.

D the balloons popping.

Question 4

In this example, the unconditioned response was

A yellow balloons.

B crying from the loud noise.

C crying at yellow balloons.

D the balloons popping.

Use the following information to answer Questions 5–7.

Ivan Pavlov was well known for his experiments involving classical conditioning. In one, he rang a bell, then presented meat powder to a dog, prompting it to salivate. With repetition, the dog began to salivate at the sound of the bell alone.

Question 5

The unconditioned stimulus in Pavlov's study was

A Pavlov.

B the meat powder.

C the sight of the bell.

D the sound of the bell.

Question 6

The dog's salivation at the sound of the bell was

A the conditioned response.

B the unconditioned response.

C the conditioned stimulus.

D both the conditioned and unconditioned response.

Question 7

Pavlov later found that while some dogs would start to salivate at the sound of different bells, and the sight of his research assistant, others would only salivate at the sound of the original bell.

The dogs that only salivated to the original bell showed

A stimulus discrimination.

B stimulus generalisation.

C spontaneous recovery.

D extinction to the original stimulus.

Question 8

Operant conditioning is based on the principle that

A behaviour with desirable consequences is likely to be repeated.

B behaviour with undesirable consequences is likely to be repeated.

C behaviour with desirable consequences is unlikely to be repeated.

D all behaviour will be repeated, regardless of the consequences.

Question 9 ●●▪

Becki's school calls her parents to inform them that she has been sending offensive text messages to her fellow students. The next day, Becki's parents take her mobile phone away from her for a week. This is most likely an example of

A positive reinforcement.

B negative reinforcement.

C positive punishment.

D negative punishment.

Question 10 ●●▪

When someone is learning to establish a response through operant conditioning, it is best to give reinforcement

A intermittently.

B every second time a response occurs.

C every time a response occurs.

D all the time, even if a response does not occur.

Question 11 ●●●

Which of the following is **not** an important consideration when administering punishment?

A Punishment must occur after the behaviour.

B Punishment must be harsh.

C Punishment must occur immediately.

D Punishment must be unwanted.

Question 12 ●●●

Which of the following is an example of a negative reinforcer?

A receiving a $200 store credit

B receiving a reduction in a jail sentence

C receiving a detention at school

D receiving a chocolate bar

Question 13 ●●▪

Which of the following must occur in both classical and operant conditioning for spontaneous recovery to take place?

A The conditioned response must be extinguished.

B There must be a rest period.

C The conditioned response must reappear.

D All of the above.

Question 14 ⬤◻◻

Ms Lanati is a high school teacher whose class is often rowdy. She decides to give each of her students a lollipop after class when they are well behaved, or a 5-minute detention after class when they are not.

According to operant conditioning, what is likely to happen under this policy?

A The class will misbehave more often.

B The class will behave well more often.

C The class will behave erratically, but then its behaviour will improve.

D There will be no change in the behaviour of the class.

Use the following information to answer Questions 15 and 16.

Hannah is a university student who has recently signed a lease to move into a rental property in an apartment block. Hannah enjoys the freedom of having her own apartment. Since moving into her new apartment, Hannah has hosted loud parties most weekends. This has resulted in her neighbours making a formal complaint to her landlord. Hannah's landlord has sent her an official letter stating that her lease will be terminated if she continues to have loud parties.

Question 15 ©VCAA 2020 SA Q14   ⬤⬤⬤

If Hannah reduces the number of loud parties in response to the contents of the letter, which principle of operant conditioning is being demonstrated?

A response cost, because Hannah's neighbours are upset by her loud parties

B negative reinforcement, because the contents of the letter achieved the desired response

C response cost, because having her lease terminated would result in Hannah losing her apartment

D negative reinforcement, because the formal complaint by the neighbours is an unpleasant consequence

Question 16 ©VCAA 2020 SA Q15   ⬤⬤⬤

Which one of the following actions is most likely to reinforce quiet and considerate behaviour from Hannah?

A the police being called when she has a loud party

B her neighbours not reacting when she has a loud party

C her neighbours waiting outside as guests arrive for a party

D being given compliments by her neighbours when she has a quiet party

Question 17 ⬤◻◻

According to social learning theory, learning is best established through which of the following types of learning?

A operant conditioning

B classical conditioning

C childhood learning

D observational learning

Question 18 ⬤⬤▪

Jason is 10 years old and learning to snowboard. He spends several hours watching video replays of the last Winter Olympics, and cannot wait to become a champion. According to observational learning principles, which component will most likely be missing to enable Jason to be able to snowboard like his idols?

A attention

B retention

C reproduction

D motivation–reinforcement

Question 19 ⬤▪▪

Albert Bandura's original experiments on aggression helped to inform observational learning theory. He found that the number of aggressive acts displayed was at its highest when observing

A a model acting pleasantly.

B a model acting aggressively.

C a model doing nothing.

D all of the above.

Question 20 ⬤▪▪

Which approach to learning is most representative of learning in Aboriginal and Torres Strait Islander cultures?

A situated multimodal systems

B behavioural

C social-cognitive

D observational learning

Section B: Short-answer questions

> **Instruction**
> Answer all questions in the spaces provided.

Question 1 (7 marks) ⬤⬤▪

What is operant conditioning? Name and explain each of the three phases of operant conditioning.

Question 2 (4 marks)

Classical and operant conditioning have many differences. Contrast each type of learning in terms of the type of response and its timing.

Question 3 (6 marks)

Many children learn by watching their parents exhibit desirable or undesirable behaviours. Outline the first three stages of observational learning, using an example of a desirable or an undesirable behaviour that a child may learn from watching their parents.

Question 4 (3 marks)

Explain the concept of situated learning as demonstrated by Aboriginal and Torres Strait Islander peoples, and explain the significance of this way of learning.

AREA OF STUDY 2
TEST 4: THE PSYCHOBIOLOGICAL PROCESS OF MEMORY

4

Section A: 20 marks. Section B: 20 marks. Total marks: 40.
Suggested time: 50 minutes

Section A: Multiple-choice questions

Instruction
For each question, circle the multiple-choice letter to indicate your answer.

Question 1

Memory is often defined as a/an _____ information-processing system.

A passive

B encoded

C active

D complex

Question 2

In terms of the information-processing system, 'encoding' refers to

A converting information into a usable form for storage.

B converting information into a usable form for retrieval.

C attending to information for storage.

D attending to information for retrieval.

Question 3

The memory store that holds a limited amount of information while it is in our conscious control is known as

A sensory memory.

B a phonological loop.

C short-term memory.

D long-term memory.

Question 4

Information about autobiographical events is stored in _____ memory, which is a branch of _____ memory.

A declarative; semantic

B semantic; declarative

C episodic; semantic

D episodic; declarative

Question 5 ⬤○○

The capacity of iconic memory is _____, and its duration is _____.

A 0.2–0.4 seconds; unlimited

B unlimited; 0.2–0.4 seconds

C 0.2–4 seconds; 5–9 items

D 5–9 items; 0.2–0.4 seconds

Question 6 ⬤○○

Renee hasn't ridden her bike for a long time but decides to use it to get fit for summer. The last time she rode her bike, she fell off, so she locked it away in the garage. After she finds her bike, she gets back on and has no trouble remembering how to ride it.

Remembering where to find the bike was a/an _____ memory, whereas remembering how to ride the bike was a _____ memory.

A semantic; procedural

B episodic; declarative

C procedural; semantic

D declarative; procedural

Question 7 ⬤○○

For information to be transferred from sensory memory to short-term memory, it must be

A attended to.

B encoded.

C rehearsed.

D retrieved.

Question 8 ⬤⬤○

In terms of the features of our memory systems, 'capacity' refers to

A how much information we can remember.

B the sort of information we can remember.

C how long we can remember information.

D how quickly we forget information.

Question 9 ⬤⬤⬤

Shantara is working very hard to remember a series of dates from the 1800s for her Politics examination by reciting them over and over. Which of the following memory stores is **least** likely to be involved?

A episodic memory

B short-term memory

C semantic memory

D declarative memory

Question 10 ○●●

Cody and Paz are studying for their upcoming surf lifesaving bronze medallion, which requires them to learn a large amount of first aid theory. The memories they are forming when learning this information are likely to be

A episodic memories.

B sensory memories.

C implicit memories.

D explicit memories.

Question 11 ●●○

Which of the following brain regions is most likely associated with procedural or task-oriented memories?

A hippocampus

B amygdala

C neocortex

D cerebellum

Question 12 ©VCAA 2018 SA Q11 ●●●

Memories that are automatic and involve how to do things, such as riding a bike, are

A explicit memories, which are stored in the cerebellum.

B implicit memories, which are processed in the cerebellum.

C explicit memories, which are consolidated by the hippocampus.

D implicit memories, which are consolidated by the hippocampus.

Use the following information to answer Questions 13–15.

Mrs Painter's Year 12 psychology class was studying furiously for the end-of-year examination. Some students, who were having trouble remembering the stages of observational learning, decided to create memory aids. Chloe put all of the initial letters together to make a word, Sophie created a sentence with the information, while Jessica attached the information to be recalled to a series of landmarks around the classroom.

Question 13 ○●●

Which of the following techniques did Chloe use?

A acronym

B acrostic

C method of loci

D narrative chaining

Question 14 ○●●

Which of the following techniques did Jessica use?

A acronym

B acrostic

C method of loci

D narrative chaining

Question 15

Which of the following is a benefit of Jessica's technique over those used by the other two girls?

A You can use it to remember large amounts of information.

B Its flexibility allows for information to be remembered in a particular order.

C You can apply the technique using physical or mental cues.

D All of the above.

Question 16

Which of the following is **not** an advantage of the use of songlines in Aboriginal cultures to aid memory? Songlines

A can feature detailed description.

B make it easy to transfer knowledge between people.

C keep the information in short-term memory.

D use the power of repetition.

Question 17

It is New Year's Eve and Madelynn is sitting on the beach picturing what the new year will look like. She closes her eyes and sees herself crossing the stage at her Year 12 graduation, then imagines what it will look like to be driving by herself on the open road as she heads off on holidays after her final exams next year.

Which memory store is Madelynn using to construct this imagined future?

A amygdala store

B short-term memory

C sensory memory

D episodic memory

Question 18

What is the name of the condition by which people cannot visualise imagery in their mind?

A autism

B aphantasia

C Alzheimer's

D aphasia

Question 19

Which area of the brain is thought to be most likely impacted in the experience of the condition referred to in Question 18?

A basal ganglia

B visual cortex

C cerebellum

D amygdala

Question 20 ©VCAA 2020 SA Q24 ●●●

Which one of the following is a characteristic of Alzheimer's disease?

A The hippocampus is the last area of the brain to be affected.

B It initially affects short-term memory more than long-term memory.

C It is caused by an increase in the level of the neurotransmitter dopamine.

D It involves neurofibrillary tangles, a build-up of abnormal protein outside the neurons in the brain.

Section B: Short-answer questions

> **Instruction**
> Answer all questions in the spaces provided.

Question 1 (10 marks) ©VCAA 2019 SB Q8 ●●●

Multi-store model of memory

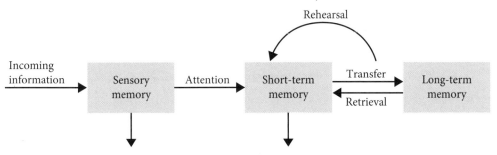

Source: adapted from P Shrestha, 'Types of Memory', in Psychestudy, 17 November 2017,
www.psychestudy.com/cognitive/memory/types

The multi-store model of memory was first proposed by Atkinson and Shiffrin (1968).
Current textbooks portray the model using a simplified diagram similar to the one above.

Discuss how the Atkinson-Shiffrin multi-store model of memory and other concepts, theories
and/or evidence can be used together to explain the formation and retrieval of the memories
of a person's first day at secondary school.

Question 2 (2 marks) ●●

Explain the role of the neocortex in memory storage, including the type of memories that it specialises in.

_____

_____

_____

_____

Question 3 (4 marks) ●●●

Explain how the use of mnemonics differs in written and oral cultures. Then give an example of a modern use of both a written and oral mnemonic.

_____

_____

_____

_____

_____

_____

_____

Question 4 (4 marks) ●●●

Discuss how episodic memories are affected in those with Alzheimer's disease. Discuss the biological factors that contribute to the onset of Alzheimer's disease, and explain what this tells us about human memory.

_____

_____

_____

_____

_____

_____

UNIT 4
HOW IS MENTAL WELLBEING SUPPORTED AND MAINTAINED?

Area of Study 1:

Test 5   The demand for sleep   28

Test 6   Importance of sleep to mental wellbeing   34

Area of Study 2:

Test 7   Defining mental wellbeing   41

Test 8   Application of a biopsychosocial approach
          to explain specific phobia   48

Test 9   Maintenance of mental wellbeing   55

Area of Study 3:

Test 10   Experimental design   62

AREA OF STUDY 1
TEST 5: THE DEMAND FOR SLEEP

Section A: 20 marks. Section B: 20 marks. Total marks: 40.
Suggested time: 50 minutes

Section A: Multiple-choice questions

Instruction
For each question, circle the multiple-choice letter to indicate your answer.

Question 1

Which of the following best describes 'consciousness'?

A an awareness of internal events only, at any given moment

B how one feels about oneself

C an awareness of both internal and external events at any given moment

D an evaluation about whether a person is alive or not

Question 2

Which of the following statements about states of consciousness is true?

A It is easy to tell if someone is in normal waking consciousness or an altered state of consciousness.

B An altered state of consciousness can be naturally or purposely induced.

C Not everyone experiences an altered state of consciousness in their lifetime.

D Consciousness is not a psychological construct.

Question 3

Which of the following is **not** an example of an altered state of consciousness?

A being drunk

B swimming

C dreaming

D meditating

Use the following information to answer Questions 4 and 5.

A person's brainwave activity can change for many reasons. Brainwaves may alter in response to concentration levels and can also change as we sleep.

Question 4 ●●○

Which brainwaves are usually exhibited by a person in normal waking consciousness?

A alpha

B beta

C theta

D delta

Question 5 ●○○

Which device can best detect, amplify and record brainwave activity?

A an EEG

B an EOG

C an EMG

D a CT scan

Question 6 ●●●

What information would an electromyograph provide for someone experiencing REM sleep?

A It would show beta-like waves.

B It would show rapid eye movement.

C It would show the participant was dreaming.

D It would show no body movement.

Question 7 ©VCAA 2018 SA Q24 ●○○

Archer's physiological responses were monitored in three separate areas, as shown in the image.

Which of the following identifies the equipment used to capture Archer's physiological responses at the points labelled 1–3 in the image?

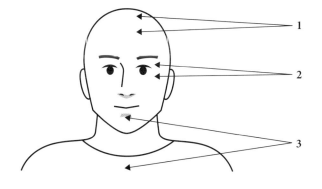

	1	2	3
A	EMG	EEG	EOG
B	EOG	EMG	EEG
C	EEG	EMG	EOG
D	EEG	EOG	EMG

Question 8 ●●○

When are alpha waves most commonly experienced?

A during NWC

B while sleepwalking

C while in NREM sleep

D while daydreaming

Question 9 `OO`

Just before you fall asleep you have slow, rolling eye movements and alpha brainwaves. This state is known as the

A REM state.

B hypnic jerk.

C hypnogogic state.

D theta state.

Question 10 `OOO`

Which of the following statements about dreams is **false**?

A Dreams mainly occur during REM sleep.

B Typically, when dreaming, the body seems as if it is in a state of paralysis.

C Usually the dreamer cannot remember the dream if woken.

D Most brainwaves when dreaming are beta-like waves.

Question 11 ©VCAA 2020 SA Q38 `O`

Which one of the following represents a quantitative measure of sleep that would indicate whether someone has moved during their sleep?

A an electroencephalograph, which indicates a deep sleep

B an electromyograph, which indicates changes in muscle tone

C a video recording, which provides visual evidence of movement

D an electro-oculograph, which records changes in eye movements indicating stages of sleep

Question 12 `O`

Which of the following best describes a circadian rhythm?

A a biological cycle that lasts for approximately 24 hours

B a biological cycle that occurs more than once in a 24-hour period

C a biological cycle that occurs less than once in a 24-hour period

D a biological cycle that lasts for 1 month

Question 13 `OO`

Which of the following cycles does **not** follow a circadian rhythm?

A sleep cycle

B body temperature

C urine production

D select hormone secretion

Question 14 `OO`

Where is the suprachiasmatic nucleus found?

A pineal gland

B hypothalamus

C hippocampus

D amygdala

Question 15 ⬤⬤⬤

Melatonin plays an important role in regulating the sleep–wake cycle. It is secreted by the _____, which is activated by the _____.

A hypothalamus; pineal gland

B amygdala; suprachiasmatic nucleus

C suprachiasmatic nucleus; pineal gland

D pineal gland; suprachiasmatic nucleus

Question 16 ⬤⬤○

Which of the following cycles follows an ultradian rhythm?

A sleep–wake cycle

B sleep cycle

C body temperature

D hormone secretion

Question 17 ⬤○○

Sarah is in her late 20s and has just given birth to her first child, Mia. Both Sarah and Mia seem to be having difficulty sleeping. On average, how many hours a day should they sleep?

A Sarah 10 hours, Mia 12 hours

B Sarah 8 hours, Mia 6 hours

C Sarah 9 hours, Mia 9 hours

D Sarah 7 hours, Mia 16 hours

Question 18 ⬤⬤○

During adolescence, about _____ hours of sleep per night is needed, and of that _____ should be spent in REM sleep.

A 9; 20%

B 6; 50%

C 9; 50%

D 6; 20%

Question 19 ⬤⬤⬤

Which of the following statements best describes a typical sleep cycle for an adult?

A The number of sleep cycles per night increases after a day of vigorous activity.

B A higher percentage of NREM sleep is experienced towards the beginning of a night's sleep.

C When first falling asleep, adults usually enter a period of REM sleep.

D On an average night, adults will complete eight sleep cycles.

Question 20 ©VCAA 2019 SA Q33 (ADAPTED) ●●

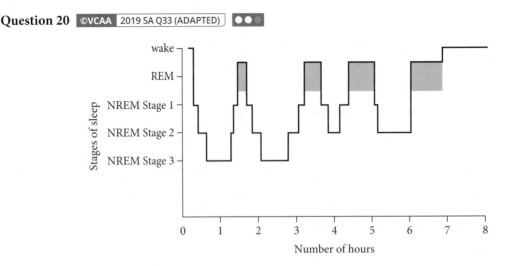

Dr Abdulla is a sleep researcher. He has collected data from four healthy participants: a child, an adolescent, an adult and an elderly person. Dr Abdulla forgot to label the hypnograms so he decided to try to identify them by considering the typical sleep patterns for each life stage.

The hypnogram shown above is likely to belong to which participant?

A the adolescent, because adolescents go to sleep later at night

B the adult, because adults have four to five sleep cycles per night

C the child, because children spend 20 per cent of their sleep in REM

D the elderly person, because elderly people wake frequently during the night

Section B: Short-answer questions

> **Instruction**
> Answer all questions in the spaces provided.

Question 1 (2 marks) ●●●

What does 'psychological construct' mean? Why is an altered state of consciousness considered a psychological construct?

Question 2 (3 marks) ●●

Discuss the features of theta waves compared to delta waves in terms of amplitude and frequency. Then describe when a person may experience delta waves.

Question 3 (11 marks) ⬤⬤⬤

a Dr TiAni is a sleep researcher who helps people with sleep problems. One of the most important things for her to track initially is how a patient's sleep progresses on a typical night. Dr TiAni has to establish when patients are in REM sleep and when they are in NREM sleep. Describe at least three ways she can distinguish between these two stages, and the devices she could use to do so.   9 marks

b Through her ongoing work, Dr TiAni notes distinct changes in the proportion of NREM compared to REM sleep from infancy to old age. Explain what she is likely to find, and why.   2 marks

Question 4 (4 marks) ⬤⬤⬤

Giorgio was involved in a scooter accident and was hospitalised for several weeks. When he returned home, his sleep–wake cycle was irregular and he had difficulty regulating his sleep–wake patterns. Give two explanations for this, and discuss the likely impact on Giorgio of his difficulty regulating his sleep–wake patterns.

AREA OF STUDY 1
TEST 6: IMPORTANCE OF SLEEP TO MENTAL WELLBEING

Section A: 20 marks. Section B: 20 marks. Total marks: 40.
Suggested time: 50 minutes

Section A: Multiple-choice questions

Instruction
For each question, circle the multiple-choice letter to indicate your answer.

Question 1

One night's total sleep deprivation can lead to which of the following effects on consciousness?

A elevated mood

B reduced cognitive abilities

C improved concentration

D difficulty completing complex tasks

Question 2

Chelsea has been getting very little sleep. Her doctor informs her that if she does not start sleeping sufficiently each night, she may suffer from symptoms associated with sleep deprivation. A physiological effect of prolonged sleep deprivation is _____, whereas a psychological effect of sleep deprivation is _____.

A poor concentration; irritability

B droopy eyelids; poor concentration

C irritability; fatigue

D fatigue; droopy eyelids

Question 3

Government advertising raises awareness about the consequences of driving while drowsy. It tells us, for example, that the effects of 24 hours of sleep deprivation on our consciousness is similar to the effects of what blood alcohol concentration?

A 0.00

B 0.01

C 0.05

D 0.10

Use the following information to answer Questions 4–6.

Sleep deprivation has a variety of causes, including shift work, jet lag or an underlying medical condition. Long-term sleep deprivation can have harmful side effects.

Question 4

Studies have shown that people deprived of REM sleep will experience _____ sleep on subsequent nights.

A longer periods of REM

B shorter periods of REM

C no REM

D only REM

Question 5

People who have gone without sleep for long periods often report having hallucinations. To reverse such effects, a person who has been deprived of a week's sleep will typically need to sleep for

A one night.

B three nights.

C 1 week.

D 2 weeks.

Question 6

Sleep-deprived people often experience very short periods of sleep while appearing to be awake. This is known as

A lack of concentration.

B hallucination.

C delusion.

D microsleep.

Question 7

Abnormalities in the typical sleep–wake cycle due to factors such as jet lag or shift work are known as

A insomnia.

B delayed sleep onset.

C sleep disorders.

D circadian phase disorders.

Question 8

Which of the following tips are **not** helpful when trying to adjust your sleep–wake cycle to prevent jet lag?

A After arrival, use natural light.

B Opt for overnight flights so you can sleep.

C Drink more coffee before a flight.

D Shift your sleep cycle gradually in the days before your flight.

Question 9 ©VCAA 2019 SA Q34 ●●●

An individual's bodily functions follow naturally occurring and predictable rhythms.

Which one of the following is true of the rhythm that individuals usually have while sleeping?

A an ultradian rhythm within a circadian rhythm

B a circadian rhythm that occurs throughout the night

C an ultradian rhythm that lasts approximately eight hours

D a circadian rhythm that matches the core body temperature rhythm

Question 10 ●○○

Bright light therapy can be used to treat circadian phase disorders. Why is the light used?

A to reset the body clock

B to wake people up

C to make the eyes sore, which leads to sleep

D to stimulate the brain

Question 11 ●○○

In delayed sleep phase disorder, people typically go to bed _____ than desired. In advanced sleep phase disorder, people typically go to bed _____ than desired.

A earlier; later

B later; later

C later; earlier

D earlier; earlier

Question 12 ●●○

Advanced sleep phase disorder (ASPD) affects only a small percentage of the population, but can be debilitating socially and professionally. Which of the following is often a complaint of ASPD sufferers?

A They can't fall asleep at night when they want to.

B They wake too early from sleep.

C They struggle to breathe while sleeping.

D They fall asleep in the morning.

Use the following information to answer Questions 13–16.

Tom is an avid gamer who not only loves playing games online, but also coaches people on how to play games online. One of Tom's teachers has recently noticed a change in his mood and alertness at school. She is concerned about his mental health, detecting symptoms of depression including extreme lethargy, lack of motivation and general low mood. She asks him about his sleep habits.

Question 13 ●●●

Why would the teacher link Tom's sleep habits to his mental health?

A Lack of sleep causes mental illness.

B Sleep hygiene is important to mental wellbeing.

C Sleep is usually a problem in adolescent males.

D She doesn't want to offend him, so asks him an easy question to get the conversation started.

Question 14 ⬛⬛⬜

Tom reports that he has not been sleeping well. He is going to bed after 2 a.m., and even though it is late, he has trouble falling asleep. He then has to get up early for school, so he is missing a lot of sleep. Which of the following pieces of advice would **not** improve his sleep health habits?

A Have a pre-bed routine.

B Go to bed at the same time each night.

C Take an afternoon nap.

D Get regular exercise.

Question 15 ⬛⬛⬛

Tom's teacher also discusses the impact of blue light before bed. Why is she worried about the impact of blue light on Tom's sleep patterns?

A Blue light increases at night from the Moon, and seeing moonlight can be bad for sleep.

B Blue light is emitted by electronic screens, and can make you addicted to the screen.

C If Tom is not seeing much natural sunlight, he will not have his required dose of blue light for the day.

D Electronic screens produce a lot of blue light, which can suppress the release of melatonin.

Question 16 ⬛⬜⬜

How soon before bed should blue light be avoided to optimise your sleep–wake cycle?

A 30 minutes to 1 hour

B 1–2 hours

C 2–3 hours

D 3–4 hours

Question 17 ⬛⬛⬜

Sleep researchers are interested in the impact of zeitgebers on the sleep–wake cycle. What are zeitgebers?

A Cues produced by your internal body clock that guide your circadian rhythm.

B Cues given by the external environment that regulate your circadian rhythm.

C Cues given by the external environment that harm the sleep–wake cycle.

D Cues given by your internal body clock that harm the sleep–wake cycle.

Question 18 ⬛⬛⬜

Which of the following is the most effective zeitgeber?

A sunlight

B eating and drinking patterns

C social interaction patterns

D exercise patterns

Question 19 ⬛⬛⬛

How does body temperature act as a zeitgeber?

A Our body temperature is on a circadian rhythm and decreases as sleep approaches.

B Our body temperature increases before bed, signalling to the body that it is bedtime.

C Our body temperature goes up and down in cycles during the day, and once we have completed eight cycles it is time for sleep.

D Body temperature is not a zeitgeber as it can't be impacted by external cues.

Question 20 ⬛⬛

Norway is known as 'the land of the midnight sun' because the sun doesn't set for more than 2 months of the year.

Which of the following statements offers the best explanation for why about one-third of Norwegians suffer from mental health problems in any given year?

A Due to the long absence of light cues, Norwegians may be more sleep deprived.

B Sleep hygiene and mental health are linked.

C Disruption to circadian rhythms can lead to negative physiological and psychological effects.

D All of the above.

Section B: Short-answer questions

> **Instruction**
> Answer all questions in the spaces provided.

Question 1 (5 marks) ⬛⬛

a In Australia, the legal blood alcohol concentration (BAC) limit to drive on a full licence is 0.05. Compare the affective and cognitive effects of someone who has a BAC of 0.05 with someone deprived of sleep for one full night.   3 marks

_____

_____

_____

_____

_____

b Should the legal drinking limit be raised to 0.10? Explain your answer using psychological evidence.   2 marks

_____

_____

_____

Question 2 (1 mark) ⬤⬤⬜

Why is delayed sleep phase disorder thought to occur during adolescence?

_____

_____

Question 3 (14 marks) ©VCAA 2020 SB Q5 ⬤⬤⬤

Predicting the body clock of shiftworkers

Australian researchers led by Doctor Julia Stone from Monash University have developed a model that makes predictions about a person's body clock based on non-invasive measurements. The usual procedure for tracking the body clock is taking blood tests that measure melatonin. In the research, in addition to measuring melatonin levels, doctors and nurses wore a wrist band that measured their physical activity and the amount of light they were exposed to over a range of day and night shifts as well as on their days off. The research demonstrated that the model provides an accurate prediction of a person's body clock shift, verified by the melatonin levels recorded.

This research could result in the development of a device that could provide shiftworkers with information about their body clock and thus help them manage the alignment of the body clock and the work schedule. Given that the impact of shiftwork can be much harder to manage than the impact of jet lag, this is welcome news.

References: JE Stone et al., 'Application of a limit-cycle oscillator model for prediction of circadian phase in rotating night shift workers', *Scientific Reports*, 9:11032, 30 July 2019, <https://doi.org/10.1038/s41598-019-47290-6>; J Elder, 'Keeping shift workers alert on the job: New study finds how to predict a person's body clock', *The New Daily*, 2 August 2019, <https://thenewdaily.com.au/life/wellbeing/2019/08/02/body-clock-shift-workers/>

a   The article uses the term 'body clock'.

Using appropriate psychological terminology, explain what the authors mean when they refer to the term 'body clock'.   2 marks

_____

_____

_____

_____

b   Why are levels of melatonin measured in this research?   3 marks

_____

_____

_____

_____

_____

_____

c Using examples, identify how the affective and behavioural functioning of doctors and/or nurses in a hospital setting may be impacted by partial sleep deprivation.   4 marks

Affective functioning

Behavioural functioning

d Explain how the device mentioned in the article could be used with bright light therapy to help doctors and nurses adjust to shiftwork in a hospital setting.   3 marks

e Why is it often much harder to realign the sleep–wake cycle in individuals doing shiftwork than in individuals experiencing jet lag?   2 marks

AREA OF STUDY 2
TEST 7: DEFINING MENTAL WELLBEING

Section A: 20 marks. Section B: 20 marks. Total marks: 40.
Suggested time: 50 minutes

Section A: Multiple-choice questions

Instruction
For each question, circle the multiple-choice letter to indicate your answer.

Question 1

After long periods of uncertainty during the COVID-19 crisis, many young people have reported that they are now better able to cope with unexpected change. This is an example of

A a mental health problem.

B increased resilience.

C decreased resilience.

D decreased stigma.

Question 2

Mental health may be considered as a continuum. Which of the following is **not** an advantage of thinking about mental health in this way?

A It normalises the experience for sufferers of mental health problems and disorders.

B It acknowledges that mental health experiences can vary.

C It acknowledges that mental health problems are often transient.

D It helps to increase stigma.

Question 3

Which of the following statements regarding suffering from a mental disorder is most accurate?

A Once diagnosed with a mental disorder, it is for life.

B Mental disorders only occur when there is external trauma.

C The severity of mental disorders can change over time.

D Mental disorders only affect adults.

Question 4

Which of the following characteristics is **not** typical of a mentally healthy person?

A low levels of anxiety

B low resilience

C strong social connectedness

D good emotional adaptability

Question 5 ⬤⬤⬤

Peta suffers from anxiety but maintains a normal job, is able to live independently and enjoys a fulfilling social life. She knows it is time to re-visit her psychologist when one of these three aspects of her life is disrupted. Thinking about mental health this way is an example of the _____ approach to viewing mental health.

A medical

B social

C historical

D functional

Question 6 ⬤⬤◯

When exploring a person's mental health and wellbeing, psychologists believe a holistic lens is important. This is because

A there is often an interplay of different factors contributing to a person's mental health.

B one psychologist alone cannot help if you are diagnosed with a mental disorder – you need lots of different perspectives.

C you are a whole person, not a series of parts.

D when you experience a mental health challenge, your whole world is turned upside down.

Question 7 ⬤⬤◯

Why is a holistic approach to wellbeing so important to Aboriginal and Torres Strait Islander peoples and their communities?

A It is important to understand an individual's context when considering mental health and wellbeing.

B There are many factors that contribute to positive mental wellbeing for Aboriginal and Torres Strait Islander peoples and their communities.

C This population group is more vulnerable to mental ill-health.

D All of the above.

Question 8 ⬤◯◯

Lily has started boarding in a school in the city. She has struggled to transition to the new environment and misses her home and community. When she discusses how she is feeling with her homeroom teacher, the teacher encourages Lily to address her mental wellbeing with physical changes, such as improving sleep, diet and exercise. Lily follows her advice, but sees little improvement.

Lily does not find her teacher's advice very helpful. Why might this be?

A Lily's teacher has applied a very narrow scope of mental wellbeing to Lily.

B Lily is not ready to get help.

C Lily did not listen to the advice.

D The homeroom teacher is not qualified to discuss mental wellbeing.

Question 9 ▢▢▢

What is the social and emotional wellbeing (SEWB) framework for Aboriginal and Torres Strait Islander peoples?

A a framework that looks at how a particular group of Aboriginal and Torres Strait Islander peoples stays mentally healthy

B a plan for making all Aboriginal and Torres Strait Islander peoples more mentally healthy

C a framework that represents a holistic approach to wellbeing for Aboriginal and Torres Strait Islander peoples, recognising the importance of culture and history in healthcare

D a framework that represents a holistic approach to wellbeing for Aboriginal and Torres Strait Islander peoples who live in urban settings

Question 10 ▢▢▢

How many domains of connection are included in the social and emotional wellbeing (SEWB) framework?

A 3

B 5

C 7

D 9

Question 11 ▢▢▢

Which of the following is **not** a domain of connection in the social and emotional wellbeing (SEWB) framework?

A connection to Country

B connection to brain

C connection to family and kinship

D connection to body

Question 12 ▢▢▢

Social and emotional wellbeing (SEWB) must be viewed through the lens of different contextual determinants of health. These include historical, cultural and social influences on Aboriginal and Torres Strait Islander peoples. Along with these three determinants, which is the fourth determinant?

A political

B medical

C psychological

D sociocultural

Question 13 ▢▢▢

Why is cultural continuity such an important cultural determinant for Aboriginal and Torres Strait Islander peoples?

A because people should not be far away from their home

B because it integrates and connects individuals and communities with the past and future of their culture

C because you can't learn about culture in other ways

D because it is the way in which values and beliefs are established in a community

Question 14

Which of the following is **not** a risk factor for the development of mental illness?

A substance abuse

B social isolation

C poor sleep

D exercise

Question 15

If phobia, stress and anxiety were placed on a mental health continuum, in which order would they appear, from healthiest to least-healthy experience?

A phobia, stress, anxiety

B anxiety, phobia, stress

C stress, anxiety, phobia

D stress, phobia, anxiety

Use the following information to answer Questions 16–18.

Lauren was very nervous on her first date with Matt. Her hands were shaky, she had trouble concentrating and she did not have much of an appetite.

Question 16

For Lauren, the date itself is best described as

A a stress system.

B a stress reaction.

C a stressor.

D stress.

Question 17

Lauren's loss of appetite is best described as

A a stressor.

B a stress reaction.

C a stress system.

D stress.

Question 18

Where would Lauren's response be placed on a mental health continuum?

A low level of mental health concern

B moderate level of mental health concern

C high level of mental health concern

D Lauren has a mental health disorder.

Use the following information to answer Questions 19 and 20.

Over a few months, Marguerite experienced significant issues with her boss at work. Marguerite is usually very optimistic and positive but her problems with her boss were making her very unhappy. She could not think of any solution. Marguerite discussed the situation with her partner, who had noticed a significant change in her attitude. Her partner suggested that she join him at the gym to help manage her stress.

Question 19 ©VCAA 2019 SA Q42 ●●●

Marguerite's current attitude would be considered a psychological risk factor because

A it is influenced by her workplace.

B it is based on her belief about the outcome.

C she is expressing her concerns to her partner.

D her emotions about the situation are caused by neurohormones.

Question 20 ©VCAA 2019 SA Q43 ●●

Which of the following identifies the internal and external factors interacting to put Marguerite's mental health at risk?

	Internal	External
A	physical health	family relationships
B	genetic predisposition to anxiety	lack of solutions
C	emotional state	interactions with her boss
D	low self-esteem	conflict resolution skills

Section B: Short-answer questions

> **Instruction**
> Answer all questions in the spaces provided.

Question 1 (5 marks) ●●

a Distinguish between the experience of good mental health and a mental disorder, using at least two contrasting examples.   4 marks

b Why is it difficult to find a definitive line to separate whether a person is mentally healthy
or not?   1 mark

_____

_____

_____

_____

_____

Question 2 (4 marks) ⬤◯◯

What is resilience? Give an example of an event that would require resilience to overcome.
Explain how an individual's resilience can alter their ability to deal with and adapt to adversity.

_____

_____

_____

_____

_____

_____

_____

_____

Question 3 (4 marks) ⬤⬤⬤

Discuss why a holistic approach to mental wellbeing, as set out by the SEWB framework
for Aboriginal and Torres Strait Islander peoples, is important for both the prevention and
treatment of mental illness. Cite a relevant principle to support your answer.

_____

_____

_____

_____

_____

_____

_____

Question 4 (7 marks) ©VCAA 2018 SB Q7

Shari moved interstate for her first job at an advertising company. She quickly found it difficult to work with the other people at the company as she considered them untrustworthy. A month after Shari started, the company underwent a restructure and Shari's job became more demanding. She struggled to meet deadlines and to think clearly. She became increasingly stressed and doubted her ability to do her job effectively. Concerned about her mental health, Shari organised an appointment with the company's psychologist.

a Identify where the psychologist might place Shari on the mental health continuum. Justify your response. 2 marks

b Describe how one relevant internal factor may have increased Shari's susceptibility to developing a mental health disorder. 2 marks

c Explain how **one** relevant psychological protective factor could influence Shari's resilience. 3 marks

AREA OF STUDY 2
TEST 8: APPLICATION OF A BIOPSYCHOSOCIAL APPROACH TO EXPLAIN SPECIFIC PHOBIA

Section A: 20 marks. Section B: 20 marks. Total marks: 40.
Suggested time: 50 minutes

Section A: Multiple-choice questions

Instruction
For each question, circle the multiple-choice letter to indicate your answer.

Question 1

An irrational, persistent and intense fear of a person, object or thing is known as

A stress.

B a phobia.

C anxiety.

D cumulative risk.

Question 2

A phobia is classified in the family of which type of mental health problems?

A mood disorder

B personality disorders

C psychotic disorders

D anxiety disorder

Question 3

The development of a specific phobia can occur for many reasons, including GABA dysfunction. What is dysfunctional about GABA in some phobia sufferers?

A There is too much.

B It fails to inhibit a stress response.

C It creates feelings of anxiety.

D It blocks neural messages from reaching the brain.

Question 4

When exploring the influence of GABA, it is important to recognise the role it plays in the body. GABA is

A an inhibitory neurotransmitter.

B an excitatory neurotransmitter.

C a sensory neuron.

D a motor neuron.

Question 5 ◼◼◻

A phobia can form through association with a negative experience but be maintained through avoidance of the negative experience. The precipitation of the phobia is learned through _____, whereas the perpetuation of the phobia is strengthened through _____.

A classical conditioning; operant conditioning

B operant conditioning; classical conditioning

C psychology; biology

D biology; psychology

Question 6 ©VCAA 2020 SA Q42 ◼◼◼

Which of the following accurately categorises both a contributing factor in the development of a specific phobia and an evidence-based intervention used to treat a specific phobia?

	Contributing factor	Evidence-based intervention
A	the stress response (biological)	exercise (social)
B	classical conditioning (biological)	challenging unrealistic thoughts (psychological)
C	catastrophic thinking (psychological)	psychoeducation (social)
D	stigma around seeking treatment (psychological)	breathing retraining (biological)

Question 7 ◼◼◻

Unfortunately, despite a change in societal perceptions, fewer men still report mental health problems than women. One of the factors that is believed to be perpetuating this problem is the impact of stigma. What is stigma?

A a positive view or image

B an avoidance strategy

C a cognitive behavioural strategy

D a negative stereotype

Question 8 ◼◼◻

What is one technique that has been shown to be effective in reducing the stigma of mental illness?

A education

B punishment

C advertising

D more phone helplines

Question 9 ◼◼◻

What effect do benzodiazepines have on the body?

A They enhance the effect of GABA to increase anxiety.

B They block the effect of GABA to reduce anxiety.

C They enhance the effect of GABA to reduce anxiety.

D They block the effect of GABA to increase anxiety.

Question 10 ⬜⬜⬜

When treating mental illness with medication, the aim for treating practitioners is to mimic, activate or inhibit the body's natural resources. Which is a true statement about benzodiazepines?

A They are an antagonist, so they bind to the receptor to activate a response.

B They are an antagonist, so they bind to the receptor to inhibit the response.

C They are an agonist, so they bind to the receptor to activate a response.

D They are an agonist, so they bind to the receptor to inhibit the response.

Question 11 ⬜⬜⬜

Some phobias evolve due to fear conditioning, which sets off neurological changes. When the neurons in the brain begin to change as a result of learning, it is known as

A GABA dysfunction.

B memory bias.

C long-term depression.

D long-term potentiation.

Question 12 ⬜⬜⬜

The kind of contributing factor to the development of a specific phobia mentioned in Question 11 is

A biological.

B psychological.

C social.

D all of the above.

Use the following information to answer Questions 13–16.

Lex is a brilliant dancer and has been learning for years. Recently, when she was performing on stage, she blanked out and forgot her routine. She has started to develop a fear of performing but doesn't want anyone to know for fear of being judged. As the months go on, her fear gets worse and she becomes convinced she has now developed a phobia of performing. She refuses to try on her costumes or practise so her teacher will be forced to not put her on stage.

Question 13 ⬜⬜⬜

The factors contributing to the development of Lex's phobia are

A dysfunction of GABA and long-term potentiation.

B specific environmental triggers and fear of stigma.

C classical conditioning and operant conditioning.

D none of the above.

Question 14 ⬜⬜⬜

Looking at the contributing factors, the most likely to be influencing the development of Lex's phobia are

A biological factors.

B psychological factors.

C social factors.

D all of the above.

Question 15

Eventually Lex's parents figure out what is going on. They sit her down and ask her how she feels. Lex tells them that she is sure that the next time she goes on stage she will forget again – then her dance troupe will lose and kick her out of the troupe, meaning she will lose all her friends and never make new ones.

Lex is exhibiting

A stigma.

B catastrophic thinking.

C stress.

D anxiety.

Question 16

Lex's parents start to challenge her assertion that if she goes on stage, it will happen again. They also get her to slow down her breathing and engage in circular breathing when she thinks about performing.

This technique is known as

A breathing retraining.

B hyperventilating.

C an anxiety control mechanism.

D cognitive behavioural therapy.

Question 17 ©VCAA 2020 SA Q49

Which of the following correctly describes cognitive behaviour therapy and systematic desensitisation?

	Cognitive behaviour therapy	Systematic desensitisation
A	requires the development of a hierarchy	requires the development of a hierarchy
B	involves the use of classical conditioning	does not involve the use of classical conditioning
C	focuses on challenging negative thought patterns	does not focus on challenging negative thought patterns
D	is not likely to involve breathing retraining	is not likely to involve breathing retraining

Question 18

Which of the following options describes the most likely limitation of using cognitive behavioural therapy to treat specific phobias?

A It can lead to long-term physical harm.

B It can lead to long-term psychological harm.

C It cannot be applied outside the controlled environment.

D It can take time and needs a willing participant.

Question 19

In systematic desensitisation, the feared stimulus is paired with a

A positive CS.

B negative UCS.

C positive UCS.

D negative CS.

Question 20 ●●▫

Systematic desensitisation is named as such because the exposure to the feared stimulus during treatment

A   decreases in intensity.

B   increases in intensity.

C   is random.

D   is rotated with stimuli that is not feared.

Section B: Short-answer questions

> **Instruction**
> Answer all questions in the spaces provided.

Question 1 (3 marks) ▫●●

What is the biopsychosocial approach? Explain how the biopsychosocial approach is used both as a preventative approach to mental illness and as a lens through which mental illness can be considered and treated.

_____

_____

_____

_____

_____

_____

Question 2 (4 marks) ●●▫

a   When comparing the experiences of anxiety and phobia, how could a phobia be classified as more dysfunctional on a mental health continuum?   2 marks

_____

_____

_____

_____

b   Discuss an appropriate social intervention in the treatment of a specific phobia. Explain how the treatment works to help a person improve their phobic response.   2 marks

_____

_____

_____

_____

Question 3 (8 marks) ©VCAA 2020 SB Q6 ●●●

When she turned four, Maxine received a medium-sized red box out of which popped a clown figure making a loud noise. When the box opened, Maxine ran away from it, towards her parents, screaming in fear. Her parents comforted her by playing with her. As a teenager, Maxine still runs away whenever she sees a similar box and her parents continue to comfort her. Maxine's parents have decided to consult a psychologist with Maxine to try to manage her phobia.

a In terms of operant conditioning, outline how Maxine's parents' response could be considered to be perpetuating her phobia of red boxes.   3 marks

_____

_____

_____

_____

_____

b The psychologist discussed with Maxine and her parents the option of using a benzodiazepine agent to manage Maxine's phobia. How could a benzodiazepine agent help manage Maxine's phobia?   3 marks

_____

_____

_____

_____

c The psychologist also indicated that memory bias may be contributing to Maxine's phobia. Explain how memory bias could be a contributing factor in Maxine's phobia.   2 marks

_____

_____

_____

_____

Question 4 (5 marks)

Beryl has a diagnosed phobia of frogs. It has grown so bad that she has even started to omit the letter 'f' from words when she speaks, because even the letter reminds her of the word, which reminds her of actual frogs.

Her phobia is really starting to interfere with her ability to function, so she seeks treatment. When discussing the origins of her phobia, Beryl tells of being a young girl growing up in northern Queensland. She says she remembers that as her family left for school each day, cane toads were everywhere, apparently staring at her. They would often jump towards her, causing extreme fear.

The treating psychologist believes that her phobia originated with classical conditioning.

a   Why does the psychologist make this assumption?   1 mark

b   Identify the unconditioned stimulus and the conditioned stimulus in this scenario.   2 marks

c   Explain how classical conditioning has led Beryl's phobia to transition from cane toads, to frogs, to the letter 'f'.   2 marks

AREA OF STUDY 2
TEST 9: MAINTENANCE OF MENTAL WELLBEING

Section A: 20 marks. Section B: 20 marks. Total marks: 40.
Suggested time: 50 minutes

Section A: Multiple-choice questions

Instruction
For each question, circle the multiple-choice letter to indicate your answer.

Question 1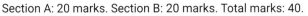

When exploring mental health, it is important to not just look at risk factors in times of need, but at protective factors as well. When can a person work to improve protective factors to help maintain their mental health?

A when they are suffering from a mental illness

B when they are recovering from a mental illness

C only when they are alrcady mentally well

D at any time

Question 2

Clay has had some struggles with anxiety, which seems to increase in intensity at times of change, like the start of a new school year or when camp approaches. He is now midway through a school year in which there will be no school camps. Clay decides to stop the cognitive behavioural therapy he had been undertaking to reduce anxiety, because there is nothing that will cause him anxiety in the near future.

Which of the following statements about Clay is true?

A Clay has made a poor decision because cognitive behavioural strategies only work when you are not suffering anxiety.

B Clay has made a poor decision because this is an excellent time to learn more about the origins of his anxiety and help put strategies in place to prevent it from reoccurring.

C Clay has made a good decision because focusing on his anxiety now will only bring it back.

D Clay has made a good decision because it will be a waste of time working to understand his anxiety when he is not suffering from it.

Question 3

Which of the following factors is most likely a biological influence on mental health?

A your parents divorcing when you are young

B feeling socially isolated

C a genetic predisposition

D having below-average intelligence

Question 4 〔⚫⚫〕

Jiayue is an excellent sleeper, and always has been. She falls asleep easily and gets a solid 9 hours every night. With the approach of Year 12 exams, Jiayue's school organises a presentation on the impact of sleep on health. Through this, Jiayue learns that

A sleep is a biological risk factor.

B sleep is a biological protective factor.

C sleep is a psychological protective factor.

D sleep is a psychological risk factor.

Use the following information to answer Questions 5–7.

Emily is an elite swimmer who is expected to be selected for the upcoming Summer Olympic games. She knows that to be selected she needs to be at the top of her game physically and mentally.

Question 5 〔⚫⚫〕

Which of the following is good advice to help Emily stay mentally well?

A Ensure you follow a good diet.

B Ensure you go on a diet.

C Ensure you train more and sleep less.

D Ensure you watch videos of your past performances, so you know where to improve.

Question 6 〔⚫⚫⚫〕

Emily knows stress can harm her mental health and self-belief. What sort of risk factor is stress?

A biological

B psychological

C social

D mental

Question 7 〔⚫⚫〕

Emily has learned that one effective way to prevent the negative impacts of stress is to engage in mindfulness meditation. Mindfulness meditation is thought to be beneficial because

A it increases patience and tolerance.

B it reduces negative emotions.

C it allows you to focus on the present.

D all of the above.

Question 8 〔⚫⚫〕

When people have preventative strategies that help their biological, psychological and social wellbeing, they are

A unable to suffer from mental health problems.

B less likely to suffer from mental health problems.

C at greater risk of suffering from mental health problems.

D no different to people without preventative strategies in their susceptibility to mental health problems.

Use the following information to answer Questions 9 and 10.

Chhaya has been a school refuser for as long as she can remember. As the start of Year 9 approaches, she is feeling optimistic and in a good place. She has several strategies that help her feel this way. When she starts to think about failing every test she completes, she remembers to adjust her thinking and remind herself that even if she does fail, there will be opportunities to resit the tests. She also practises arriving at school, getting out of the car and heading to her locker the week before school starts.

Question 9 〔◖◦◦〕

Chhaya adjusting her thinking is an example of a _____ strategy; Chhaya practising getting out of the car each day is a _____ strategy.

A social; biological

B psychological; biological

C cognitive; behavioural

D behavioural; social

Question 10 〔◦◦◦〕

Chhaya's ability to readjust her thinking is an example of

A meditation.

B resilience.

C catastrophic thinking.

D reframing.

Question 11 〔◦◦◦〕

Having a support network of family and friends with whom you feel comfortable sharing your feelings is an important protective strategy for mental health. Sometimes people can be concerned about sharing their feelings with others for fear of being labelled. This can be known as

A stress.

B rumination.

C dissociation.

D stigma.

Use the following information to answer Questions 12 and 13.

Peter moved from London when he was 12 years old to live in Australia. He struggled with the school transition and found it very hard to adapt to Australian culture. He did, however, forge new friendships easily.

Question 12 〔◦◦◦〕

Peter's friendships may have been a _____ protective factor contributing to his mental wellbeing.

A psychological

B social

C biological

D spiritual

Question 13 ⬤⬤⬤

As time goes on, Peter finds that his friendships are somewhat superficial. He thinks people were probably drawn to him because he was new and different, and realises he doesn't have any common interests with his friends. He also feels as though his friends are being nice to him to gain the respect of the teachers. Peter's social connections are probably lacking

A authenticity.

B past connection.

C stigma.

D resilience.

Question 14 ⬤⬤⬤

What does 'cultural determinant' mean, when exploring maintenance of wellbeing for Aboriginal and Torres Strait Islander peoples?

A the culture that someone belongs to

B whether or not you will experience mental illness

C the extent to which cultural contexts enable people to learn about and express their culture

D the extent to which culture determines the person you become

Question 15 ⬤⬤⬤

Which of the following is an example of a cultural determinant?

A cultural belonging

B belonging to Country

C self-destruction

D cultural continuity

Question 16 ⬤⬤⬤

For Aboriginal and Torres Strait Islander peoples, transgenerational trauma and racism are two experiences that

A may negatively impact mental wellbeing.

B may positively impact mental wellbeing.

C are biological approaches to mental wellbeing.

D have no effect on mental wellbeing.

Question 17 ⬤⬤⬤

Spirituality is extremely important in Aboriginal and Torres Strait Islander cultures to maintaining mental wellbeing. What is spirituality?

A our sense of connection to something bigger than ourselves, shaping our identity

B our sense of connection to the mind

C our sense of connection to land

D our sense of connection to our community

Question 18 `OO`

The Uluru Statement from the Heart articulates the importance of including an Aboriginal and Torres Strait Islander voice in Parliament. This is a power application of

A cultural continuity.

B self-determination.

C contextual determinants.

D political continuity.

Question 19 `OO`

How is information about maintaining mental wellbeing traditionally kept in Aboriginal and Torres Strait Islander cultures?

A It is published in journals.

B It is documented in texts.

C It is passed from generation to generation.

D It is stored in the community in which it was developed.

Question 20 `O`

Which of the following is a key benefit of the way information about mental wellbeing is shared in Aboriginal and Torres Strait Islander cultures?

A Everyone in the community gets to become a healer.

B Cultural continuity is maintained.

C The information is not specific to communities so it can be applied effectively and broadly.

D No one ever suffers from mental health problems.

Section B: Short-answer questions

> **Instruction**
> Answer all questions in the spaces provided.

Question 1 (2 marks) `OO`

Does a genetic predisposition to being mentally well mean that a person will never suffer from a mental illness? Explain your answer.

Question 2 (3 marks) ⬤⬤⬜

What is the biopsychosocial approach? Describe one advantage and one disadvantage to viewing mental health in this way.

_____

_____

_____

_____

_____

_____

Question 3 (6 marks) ⬤⬜⬜

Since the death of his wife last year, Terry has had difficulty sleeping. He is drinking a lot of coffee to help keep him awake during the day, but he is lethargic and unmotivated.

He decides to take a holiday to refocus. During this time he stays away from coffee and alcohol and improves his diet. He attends regular guided meditation sessions at the resort he is staying at, which he aims to continue at home. He also uses the time away to chat to friends, and arranges to connect with them after he comes home, hoping to fill some of the space left by his wife's absence.

Terry has no doubt that he will have many bad days in the years to come, but feels he is putting things in place to improve his ability to cope with the inevitable challenges. Discuss how each of the following factors (biological, psychological and social) may help Terry maintain his mental wellbeing in the years to come. Support your answer with relevant examples.

Biological factors _____ 2 marks

_____

_____

_____

Psychological factors _____ 2 marks

_____

_____

_____

Social factors _____ 2 marks

_____

_____

_____

Question 4 (9 marks) ⚫⚫⚫

Name and explain two cultural determinants that contribute to mental wellbeing in Aboriginal and Torres Strait Islander cultures and outline the ways in which they help. Explain how these cultural determinants are connected, and the ways in which they are different.

_____

_____

_____

_____

_____

_____

_____

_____

_____

_____

_____

_____

AREA OF STUDY 3
TEST 10: EXPERIMENTAL DESIGN

10

Section A: 20 marks. Section B: 20 marks. Total marks: 40.
Suggested time: 50 minutes

Section A: Multiple-choice questions

Instruction
For each question, circle the multiple-choice letter to indicate your answer.

Use the following information to answer Questions 1–3.

Mitchell is conducting research to establish whether meditating for an hour a day reduces reported stress levels. He divides research participants into two groups to test this. Group A completes 1 hour of meditation by listening to a podcast just before going to bed for 1 week, while Group B is instructed not to undertake any meditation practices for a week.

At the beginning of the study, participants are asked to rate their stress level on a scale of 1–10. This is repeated at the conclusion of the experiment.

Question 1 ●●■

What was the independent variable for this study?

A the difference in the stress rating reported before and after the experiment

B the results after the experiment

C the presence of meditation

D the time of day meditation occurred

Question 2 ●●■

What was the dependent variable for this study?

A the difference in the stress rating reported before and after the experiment

B the results after the experiment

C the presence of meditation

D the time of day meditation occurred

Question 3 ●●●

Mitchell asks participants in both groups to take annual leave for the week of the experiment to try to eliminate differences in stressors during the week. The week of no work is known as

A an extraneous variable.

B an outlier.

C a confounding variable.

D a controlled variable.

Question 4

In psychological research, the term 'random allocation' refers to

A every member of the sample having an equal chance of being in the control or experimental group.

B every member of the population having an equal chance of being in the sample.

C every member of the population having an equal chance of being in the control or experimental group.

D every member of the sample having an equal chance of being in the population.

Question 5

To eliminate extraneous variables caused by participant expectations and any problems associated with the experimenter being aware of who is in the control and experimental groups, researchers can employ a _____ procedure.

A single-blind

B double-blind

C double-placebo

D Hawthorne

Question 6

Which of the following sampling techniques gives researchers the best chance of obtaining a sample that is representative of the population?

A random sampling

B stratified sampling

C convenience sampling

D random allocation

Use the following information to answer Questions 7–9.

Samantha is conducting a study for Glenvale Hospital and wants to see if there is a link between time spent in hospital and the onset of depression. She divides the patients in the hospital into age groups, then selects 40 people to be part of her study.

Question 7

The population for study is _____, while the sample is _____.

A the 40 patients in the study; all patients at the hospital

B all patients at the hospital; the 40 patients in the study

C all people living in Victoria; all patients staying at the hospital

D all people in hospitals in Victoria; all patients at Glenvale Hospital

Question 8 〔●●■〕

Samantha's research assistant suggests she compares the results of two different groups: those who spent 2 days in the hospital (control group) and those who spent 1 week in the hospital (experimental group).

Why is it important to have a control group when conducting research?

A to obtain more results

B to test whether the dependent variable causes a change in the independent variable

C to determine the effect of the independent variable by comparing the results of the control group with the results of the experimental group

D to eliminate extraneous variables due to differences in participant characteristics

Question 9 〔●●■〕

When looking at the results, Samantha notices that some participants in the study already had a diagnosis of depression before their hospital admission. Samantha had not controlled for this in the study, and as a result this may be

A a confounding variable.

B an independent variable.

C a dependent variable.

D a placebo effect.

Use the following information to answer Questions 10 and 11.

Albert Bandura was well known for his work on social learning theory. He had many variations of experiments that measured the effect of observational learning on aggression.

One of his experiments investigated whether watching different types of models – such as cartoon models or live models – would influence the number of aggressive acts exhibited by children.

Question 10 〔●●■〕

In Bandura's experiment, what was the independent variable?

A the number of children included

B the number of aggressive acts exhibited

C the intelligence of the children

D the type of model children viewed

Question 11 〔●●■〕

In Bandura's research, it was found that boys had a higher mean number of aggressive acts than girls. How is the mean calculated?

A by finding the mid-point in a group of scores

B by finding the most commonly occurring number in a group of scores

C by subtracting the lowest from the highest score

D by adding up all the scores and dividing the total by the number of scores

Use the following information to answer Questions 12 and 13.

For his PhD thesis, Rick decides to test the effectiveness of physical exercise as stress relief on two groups of people. He first asks participants to report on their current stress level and to score it out of 10. He then divides participants into two groups, ensuring that both groups are equal in terms of their reported stress level, and that there is an equal number of males and females in each group.

He asks one group of participants to complete some exercise then report on their stress levels. The other participants complete no exercise, and report on their stress levels an hour later. Rick then compares the two stress scores.

Question 12

The research design that Rick is most likely employing is

A a matched-participants design.

B a repeated-measures design.

C an experimental design.

D an independent-groups design.

Question 13

When planning his research, Rick considers using other experimental designs, such as using the same group of participants in each group and testing them on two different days. Rick decides the disadvantage of this design is that it may be influenced by

A counterbalancing.

B single-blind effects.

C placebo effects.

D order effects.

Question 14

One of the greatest advantages of using a literature review as a research methodology is that it utilises which type of data?

A primary data

B secondary data

C subjective data

D qualitative data

Question 15

What is one advantage of using qualitative data as a means of data collection?

A It is easy to compare pieces of data.

B It is on a numerical scale.

C It gives participants a chance to explain their behaviour.

D It is quick to collect.

Question 16 ⬤⬤⬤

Which measure of central tendency is **least** affected by outliers?

A median

B mode

C mean

D variance

Question 17 ⬤⬤⬤

When evaluating the potential source of error, it is important that the research findings can be replicated with different samples and in different environments. This means that the research would have high _____.

A variance

B validity

C reliability

D uncertainty

Question 18 ⬤⬤⬤

Alan has signed up to be part of a psychological experiment that involves assessing which areas of his brain are used when he performs simple mathematical equations. When Alan is debriefed at the conclusion of the study, he is told he has achieved quite a low score on the maths test and had very low levels of activity in his brain when attempting these equations.

Alan is quite embarrassed by his low score and asks for his results not to be used in the study. According to ethical guidelines, Alan is

A not allowed to remove his results as the study has already been completed.

B allowed to do this, but it is not an ethical right.

C exercising his right to informed consent.

D exercising his withdrawal rights.

Question 19 ⬤⬤◯

Dr Jenson advertises in a local paper for participants to be involved in research based on neural functioning, and 50 people respond. These participants have exercised their right to

A informed consent.

B voluntary participation.

C confidentiality.

D withdrawal rights.

Question 20 ⬤⬤◯

Which of the following is **not** a purpose of debriefing when conducting psychological research?

A to help alleviate any psychological harm

B to allow participants to obtain information about their individual results

C to enable researchers to meet their participants and to decide whether to use their results or not

D to allow participants to ask questions about the research

Section B: Short-answer questions

> **Instruction**
> Answer all questions in the spaces provided.

Question 1 (10 marks)

Dr Pham is conducting research into the most effective form of parenting in the United States. She advertises for a group of parents in Texas to be involved in her study. The 50 parents chosen have children aged 2–15.

Dr Pham asks half the parents to only use positive behaviours with their children: to praise them when they do the right thing, and not punish poor behaviour at all. She asks the other group of parents to use no positive behaviours when their children are well behaved, but to always punish poor behaviour by placing the child in 'time out'.

After a month, Dr Pham asks the parents to record the number of negative behaviours seen in their children in the final 3 days of the study, so she can determine whether reward or punishment is more effective when parenting.

Dr Pham finds that those who use reward record fewer negative behaviours than those who use punishment.

a Identify the independent and dependent variables in this experiment.

Independent variable _____ 1 mark

_____

Dependent variable _____ 1 mark

_____

b Write a possible hypothesis that may have been tested.   2 marks

_____

_____

_____

_____

c   Write a section of a discussion for this research containing:
   - an explanation of at least two extraneous variables and the impact they may have on the results
   - future improvements that address these research flaws.   6 marks

Question 2 (10 marks) ⬤⬤⬤

A new researcher believes she can study the changes in neurons when a memory is formed by using a new brain imaging device. She would like to conduct her research on children aged 5–15 years enrolled in state schools in St Kilda, Melbourne. Because she wants her research to have few design flaws, she would like to ensure that she selects a sample in a way that reduces as many extraneous variables as possible. As this is new research, she would like to start by using a case study on the 10 students she selects.

Design the experimental procedures and considerations for this new research. Your answer should include:

- the aim for this experiment
- an explanation of the most appropriate sampling technique and how the sample could be selected, as well as discussion about the types of research flaws this sampling technique eliminates
- an explanation of the advantages and disadvantages for using a case study for this work
- a discussion of ethical considerations of particular relevance to children, and how these considerations will be adhered to.

EXTRA WORKING SPACE

STUDENT NUMBER

Letter

PSYCHOLOGY
Written examination

Reading time: 15 minutes
Writing time: 2 hours 30 minutes

QUESTION AND ANSWER BOOK
Structure of book

Section	Number of questions	Number of questions to be answered	Marks
A	50	50	50
B	13	13	70
		Total	120

- Students are permitted to bring into the examination room: pens, pencils, highlighters, erasers, sharpeners and rulers.
- Students are NOT permitted to bring into the examination room: blank sheets of paper and/or correction fluid/tape.
- No calculator is allowed in this examination.

Materials supplied
- Question and answer book of 22 pages
- Additional space is available at the end of the book if you need extra space to complete an answer.

Instructions
- Write your **student number** in the space provided above on this page.
- Check that your **name** and **student number** as printed on your answer sheet for multiple-choice questions are correct, **and** sign your name in the space provided to verify this.
- Unless otherwise indicated, the diagrams in this book are not drawn to scale.
- All written responses must be in English.

Students are NOT permitted to bring mobile phones and/or any other unauthorised electronic devices into the examination room.

© VICTORIAN CURRICULUM AND ASSESSMENT AUTHORITY

Section A: Multiple-choice questions

Question 1

When experiencing heightened arousal, which of the following combination of events occurs?

A   mouth dries, blood drains from face, heart rate increases, digestion is stimulated

B   pupils dilate, mouth dries, digestion is inhibited, heart rate increases

C   breathing rate increases, pupils contract, palms sweat, digestion is inhibited

D   bladder relaxes, bronchioles contract, palms sweat, pupils dilate

Question 2 ©VCAA 2020 SA Q2

Which statement about conscious or unconscious responses by the nervous system is correct?

A   A conscious response by the nervous system is involuntary and goal-directed.

B   A conscious response by the nervous system is voluntary and attention is given to the stimulus.

C   An unconscious response by the nervous system is voluntary and regulated by the autonomic nervous system.

D   An unconscious response by the nervous system is unintentional and is always regulated by the autonomic nervous system.

Question 3

Which of the following neurotransmitters has an inhibitory effect?

A   gamma-aminobutyric acid (GABA)

B   epinephrine

C   norepinephrine

D   all of the above

Question 4

Many people gamble occasionally for entertainment, but some become addicted. Which neurotransmitter is associated with addictive behaviours such as problem gambling?

A   glutamate

B   GABA

C   serotonin

D   dopamine

Question 5

The natural plasticity of the brain to adapt to new experiences as we age is known as

A   developmental plasticity.

B   adaptive plasticity.

C   survival plasticity.

D   a critical period.

Question 6 〔●●●〕

During the fight–flight–freeze response, which of the following physiological responses is least likely to occur?

A   pupils contract

B   perspiration increases

C   saliva production decreases

D   heart rate increases

Question 7 〔●●●〕

In the first months of the year, Sarah complained of headaches and of always feeling tense and tired. Since then, little has changed, apart from her workload increasing. Yet she now says that she feels 'OK'. Sarah is most likely

A   in the shock stage of the General Adaptation Syndrome.

B   in the resistance stage of the General Adaptation Syndrome.

C   in the exhaustion stage of the General Adaptation Syndrome.

D   having a nervous breakdown.

Question 8 〔●●●〕

During the stage you identified in Question 7, the body's normal level of resistance to stress

A   is below normal.

B   is above normal.

C   is initially above normal but rapidly drops below.

D   fluctuates above and below normal.

Question 9 〔●●●〕

Which of the following best describes the role of cortisol in chronic stress?

A   It suppresses the stress response.

B   It makes a person feel stressed.

C   It supports changes in the body that help it respond to stress.

D   It eliminates adrenaline.

Question 10 〔●●●〕

Thania has just lost her part-time job at a jewellery store. She didn't really like the job and had been thinking about doing something different anyway, so she doesn't know how she feels about the news. According to Lazarus and Folkman's Transactional Model of Stress and Coping, which stage is Thania most likely in?

A   shock

B   countershock

C   primary appraisal

D   secondary appraisal

Question 11 ©VCAA 2019 SA Q7 ●●●

One limitation of Lazarus and Folkman's Transactional Model of Stress and Coping is that the model

A   fails to explain the outcome if coping resources are inadequate.

B   does not account for the different interpretations of events by individuals.

C   does not recognise that the individual and the environment both play a role in the stress response.

D   is unable to be researched experimentally because primary and secondary appraisals often occur simultaneously.

Question 12 ●●●

Omah's homework is mounting up and he just doesn't know where to begin. He decides that it is easier to leave it all until the weekend and tackle it then, right before the deadline.

Which strategy for coping with stress is Omah using?

A   avoidance strategy

B   meditation

C   approach strategy

D   exercise

Question 13 ●●●

When developing an effective strategy for coping with stress, which of the following is an important consideration?

A   The strategy is difficult to use.

B   The strategy works in a controlled, stress-free setting.

C   The strategy is rarely practised.

D   The strategy is appropriate in the necessary context.

Use the following information to answer Questions 14–16.

Ros has a beautiful holiday house in Portsea that she visits on weekends. While she is there, she loves to listen to CDs from her old collection while cooking dinner. One day, Ros realises that when she listens to the particular CD she always starts with, she salivates.

Question 14 ●●●

Ros' response to the CD is an example of

A   observational learning.

B   shaping.

C   classical conditioning.

D   operant conditioning.

Question 15 ●●●

The songs on the CD are the

A   conditioned stimulus.

B   unconditioned stimulus.

C   conditioned response.

D   unconditioned response.

Question 16 [●●●]

Ros downloads a live recording of the music she always listens to when she starts cooking. When she plays it, she does not salivate. Why not?

A   She isn't hungry.

B   She is experiencing stimulus generalisation.

C   She is experiencing stimulus discrimination.

D   She is experiencing spontaneous recovery.

Question 17 [●●●]

In classical conditioning, the learner is _____, while in operant conditioning, the learner is _____.

A   voluntary; involuntary

B   involuntary; voluntary

C   active; passive

D   passive; active

Question 18 ©VCAA 2019 SA Q15 [●●●]

Five-year-old Frank does not put his rubbish in the bin even though he has watched his parents do so many times. He has also reminded his younger sister to put her rubbish in the bin.

In terms of observational learning, Frank will be most likely to put his rubbish in the bin if he

A   has the motivation to put rubbish in the bin.

B   is developmentally ready to put rubbish in the bin.

C   pays more careful attention to his parents' behaviour.

D   develops a mental representation of putting rubbish in the bin.

Question 19 [●●●]

Albert Bandura was known for his work on observational learning. Which of the following findings would most likely support his social learning theory?

A   Children who viewed a model being punished showed significantly more aggressive acts than those who saw a model receive no consequences.

B   Children who viewed a model being punished showed fewer aggressive acts than those who saw a model receive no consequences.

C   Children who viewed a model being punished showed more aggressive acts than those who saw a model being rewarded.

D   All of the above.

Question 20 [●●●]

To gauge the effect of observational learning, Bandura measured the mean number of aggressive acts. The mean is a measure of central tendency. Which of the following statistical measures is also a measure of central tendency?

A   variance

B   standard deviation

C   reliability

D   median

Question 21 ©VCAA 2020 SA Q39 ●●○

The mean is a measure of central tendency often used in psychological research.

The mean can be a misleading representation of data if

A   the frequency of each score has not been calculated.

B   the range of the scores is much greater than anticipated.

C   it contains outliers, very small or large values in the scores that are not typical.

D   there is not an equal number of scores whose values lie above and below its value.

Question 22 ●●●

While there are many forms of research methodology, one of the best ways to obtain primary data is to conduct an experiment. An experiment seeks to measure

A   the behaviour of a small group of people.

B   whether a cause-and-effect relationship exists between two variables.

C   whether certain behaviours are evident in their natural environment.

D   whether correlations can be drawn between two different pieces of research.

Question 23 ●●○

Research has shown that the capacity of short-term memory is _____, while the duration of short-term memory is _____.

A   18–20 items; 5–9 seconds

B   18–20 seconds; 5–9 items

C   5–9 seconds; 18–20 items

D   5–9 items; 18–20 seconds

Question 24 ●●●

Which of the following brain regions is most likely associated with storing implicit memories?

A   cerebellum

B   amygdala

C   neocortex

D   hippocampus

Question 25 ●○○

In the study of music, the mnemonic 'Every Good Boy Deserves Fruit' is an example of

A   an acronym.

B   an acrostic.

C   the method of loci.

D   a songline.

Question 26

Which of the following is a mnemonic that is heavily utilised in oral cultures?

A   songlines

B   acronyms

C   acrostics

D   method of loci

Question 27

Which of the following statements regarding REM sleep is **false**?

A   REM sleep is often known as paradoxical sleep.

B   During REM sleep, the body is in a state of paralysis.

C   Dreaming usually occurs during REM sleep.

D   REM sleep decreases in duration throughout the night.

Question 28

Why are babies thought to have a decreased proportion of REM sleep in their sleep cycle as they transition to childhood?

A   Their rate of neural development increases.

B   Their physical development slows down.

C   Their neural development slows down.

D   Their physical development increases.

Use the following information to answer Questions 29–32.

Phoenix and her friends drove to a music festival in Byron Bay. They arrived the night before the festival and they all slept in Phoenix's small car for the night. Phoenix and her friends all experienced very disturbed sleep.

Question 29 ©VCAA 2019 SA Q29

What behavioural effect may Phoenix and her friends experience the next day due to being partially sleep-deprived?

A   an inability to sit still while listening to the music

B   a lack of interest in making conversation with each other

C   being unable to remember the names of all the bands that they were listening to

D   feeling particularly hungry and wanting to visit a food truck for burgers and chips

Question 30 ©VCAA 2019 SA Q30

At one point Phoenix was unable to remember the hair colour of the lead singer of her favourite band. Sleep deprivation is likely to contribute to her poor memory because

A   sleep deprivation can result in poor cognitive functioning.

B   affective functioning is compromised by sleep deprivation.

C   music festivals have a compounding effect on sleep deprivation.

D   the hallucinatory effects of sleep deprivation will cause memory problems.

Question 31 ©VCAA 2019 SA Q31 ●●●

Phoenix and her friends stayed up watching bands all night. The next morning, Phoenix wanted to drive her car home, despite having not slept for the entire previous night or day. Her friends urged her not to drive because of the effect that sleep deprivation may have on her concentration.

The most accurate information to support the concern of Phoenix's friends is that a full night's sleep deprivation is equivalent to a blood alcohol concentration (BAC) of

A   0.10 and her eyelids might droop.

B   0.05 and she might have slower reaction times.

C   0.10 and she might not stay within her lane on the road.

D   0.05 and she might find it difficult to maintain the speed at which she is travelling.

Question 32 ©VCAA 2019 SA Q32 ●●●

In her sleep-deprived state, Phoenix's brain waves would be similar to those of an individual who has been administered a depressant.

Compared to her consciousness when she began driving to the music festival, by the third day, Phoenix would most likely have

A   lower levels of alertness and more beta waves.

B   higher levels of alertness and more alpha waves.

C   higher amplitude and lower frequency brain waves.

D   lower amplitude brain waves and fewer beta waves.

Question 33 ●●●

Inability to regulate the sleep–wake cycle may be due to damage to the

A   pineal gland.

B   suprachiasmatic nucleus.

C   hypothalamus.

D   any of the above.

Question 34 ●●○

Sufferers of advanced sleep phase disorder typically have disruption to their

A   ultradian cycle.

B   circadian cycle.

C   delayed sleep cycle.

D   all of the above.

Question 35 ●●●

People experiencing difficulty sleeping may visit a sleep laboratory to measure different physiological responses. One advantage of gathering data on physiological responses is that the data collected is

A   objective.

B   subjective.

C   qualitative.

D   artificial.

Use the following information to answer Questions 36 and 37.

Hope enjoys reading before going to sleep. Every night, once in bed, she will pick up her book and read for 15 minutes. She will then put on her night-time face cream that smells of lavender and turn out the light.

Question 36

Consider what Hope does before she goes to bed. It is likely she is making the most of

A   sleep starters.

B   zeitgebers.

C   sleep hygiene.

D   circadian clocks.

Question 37

Which piece of advice could Hope benefit from to further improve the onset of sleep and hence the sleep–wake cycle?

A   Limit screen time for 30 minutes before sleep.

B   Go to bed whenever you start to feel tired.

C   Go to bed at the same time each night.

D   Sleep with a night light on.

Question 38

Melissa is 30 and lives with her boyfriend. Her father has just died. Her extended family is under significant financial pressure as a result, and she is having to help support them, both emotionally and financially.

In terms of the strain on her mental health, the death of her father is an _____ factor, and the financial strain is an _____ factor.

A   external; external

B   external; internal

C   internal; internal

D   internal; external

Question 39

Mentally healthy people can deal with most of life's stressors when they come their way. This is known as

A   resistance.

B   resilience.

C   eustress.

D   de-stress.

Question 40

What is one similarity between anxiety and a phobia?

A   They are only experienced in adulthood.

B   Neither are mental illnesses.

C   Neither can be treated.

D   They both involve feelings of worry.

Question 41 ©VCAA 2018 SA Q37 ●●●

Anxiety can be distinguished from phobia because only anxiety

A   involves distress.

B   can be helpful in mild amounts.

C   triggers the fight–flight–freeze response.

D   is influenced by biological, psychological and social factors.

Question 42 ●●●

Which of the following is **not** a psychological factor that contributes to the development of a specific phobia?

A   cognitive bias

B   catastrophic thinking

C   environmental triggers

D   memory bias

Use the following information to answer Questions 43–45.

Theodore lost his job two years before he intended to retire and it had a negative impact on his mood and ability to cope. He did not pay two electricity bills despite having sufficient funds. He became withdrawn while at his golf club and soon stopped playing. When he also started complaining of sleeping problems, his daughter encouraged him to see his family doctor with her.

Question 43 ©VCAA 2020 SA Q43 ●●

Theodore might be showing signs of a mental health problem because

A   not paying bills indicates a mental health problem.

B   both internal and external factors are contributing to his behaviour.

C   internal factors rather than external factors are influencing his behaviour.

D   his behaviour is uncharacteristic and has had a negative impact on his wellbeing.

Question 44 ©VCAA 2020 SA Q44 ●●●

To try to improve Theodore's resilience, the doctor decided to focus on biological, psychological and social factors. Which of the following is a combination of the most likely strategies the doctor would initially encourage or use?

	Biological	**Psychological**	**Social**
A	prescribing medication	challenging Theodore's negative thought patterns	reminding Theodore about his supportive family
B	getting Theodore's friends to bring him meals	getting Theodore to join the seniors' club	getting Theodore to start an exercise program
C	organising genetic testing	getting Theodore to make lists of the things he needs to do	taking Theodore to a nutritionist
D	organising nutritious meals	challenging Theodore's negative thinking about the future	getting Theodore's friends to visit him regularly

Question 45 ©VCAA 2020 SA Q45 ●●●

Theodore's unemployment was making him feel stressed. He applied for jobs advertised in the newspaper, but this did not result in employment. He went to see a careers counsellor, who suggested some further strategies for finding new employment.

Which one of the following statements applies to the coping strategy used by Theodore when he saw a careers counsellor?

A   It is an approach strategy that does not demonstrate coping flexibility.

B   It has context-specific effectiveness and demonstrates coping flexibility.

C   It has context-specific effectiveness and will help him avoid stressful situations.

D   Given the difficulty older people have finding a job, it demonstrates coping inflexibility.

Question 46 ©VCAA 2018 SA Q36 ●●●

A researcher was investigating the effects of a gamma-aminobutyric acid (GABA) agonist in the treatment of a specific phobia. Group A, the experimental group, received the GABA agonist. Group B, the control group, received a placebo. Concerned about experimenter bias, the researcher used a double-blind procedure with the help of a research assistant who worked directly with the participants.

Which one of the following identifies the double-blind procedure used in this investigation?

A   Only the researcher knew who would receive the placebo.

B   Only the research assistant knew who would receive the GABA agonist.

C   Only the researcher and the control group knew who would receive the placebo.

D   Only the researcher and the research assistant knew who was in the experimental group and the control group.

Question 47 ●●●

When conducting research, it is important to ensure that the research is actually measuring what it claims to. In mental health research, for example, it is difficult to determine differences between anxiety symptoms and the normal experience of stress. To ensure that treatments effectively improve anxiety and that the results are accurate, researchers must examine and address this issue.

In psychological research, what is this known as?

A   outliers

B   reliability

C   validity

D   ethics

Question 48 ●●●

Which ethical consideration ensures that research participants' data remains private and is not disclosed?

A   confidentiality

B   deception

C   debriefing

D   informed consent

Question 49 ©VCAA 2018 SA Q35 ●●●

Which of the following would most assist researchers with minimising extraneous variables in an experiment?

A   single-blind procedures and use of a placebo

B   convenience sampling and standardised procedures

C   standardised instructions and double-blind procedures

D   counterbalancing to control order effects and experimenter bias

Question 50 ©VCAA 2020 SA Q50 ●●●

One of the strengths of using secondary data from the internet for psychological research is that secondary sources will

A   have satisfied ethical guidelines.

B   have already been published and so the data is likely to be reliable and valid.

C   provide large reserves of data and be representative of the general population.

D   provide access to volumes of data that the researcher may not be able to gather.

Section B

Instructions for Section B
Answer all questions in the spaces provided.

Question 1 (7 marks) ▢▢▢

a John was sitting in the car waiting for his friend. It was a hot day so he opened the window and rested his arm on the door. Suddenly, a mosquito landed on his arm and he moved his other hand to swat it.

Write a detailed description of the role played by John's peripheral and central nervous systems when he felt the bug on his arm and then reacted to it, considering it was not an automatic reaction to swat the bug.   4 marks

b How would John's neural response differ if instead of a mosquito, an open flame touched his arm?   3 marks

Question 2 (4 marks) ▢▢▢

a Explain what long-term potentiation (LTP) is and give an example of a physiological change that occurs in forming a consolidated memory.   2 marks

b Explain how LTP relates to developmental plasticity and why LTP is important.   2 marks

End of Question 2

Question 3 (3 marks) ◐○○

Nassim loves writing to-do lists at the start of her week, then ticking off the items once they are achieved. Sometimes she even writes tasks on the list just so she can tick them off. With reference to neurotransmitters, explain why Nassim enjoys ticking things off her to-do list.

Question 4 (2 marks) ◐◐○

Explain what is meant by the 'bidirectional nature of the gut–brain axis'.

Question 5 (8 marks) ©VCAA 2019 SB Q3 ◐◐◐

SNAKE AVOIDANCE TRAINING SAVES DOGS' LIVES
by Kwan Pelucci

Veterinary surgeon Margie Grey is a strong advocate of training dogs to avoid snakes. 'Snake avoidance training saves dogs' lives', said Dr Grey. To prevent dogs from being bitten, she regularly engages an animal trainer specialising in reptiles to work with her clients and their dogs. The methods the trainer uses are described below.

During the training sessions, the dogs wear a collar that delivers a low-level electric shock sent by a remote transmitter held by the animal trainer.

Over two or more sessions, each dog is exposed to a range of non-venomous snakes and trained
to avoid these snakes. The animal trainer uses two different training methods in each session.

The first method involves conditioning the dogs to associate snakes with the electric shock delivered through their collar.

The second method involves giving the dogs a treat each time the dogs choose not to approach a snake.

Dr Grey said that even though the first part of the training uses an electric shock, she trained her own dogs using this method because she knows the treatment saves dogs' lives.

Question 5 continues on page 85

a   In terms of classical conditioning, describe how the animal trainer creates an association between the neutral stimulus and the unconditioned stimulus to develop a conditioned response.   3 marks

_____

_____

_____

_____

_____

_____

b   Operant conditioning was also used in this training.

Name the antecedent, the subsequent behaviour and the type of consequence in the training sessions.   3 marks

Antecedent _____

_____

Behaviour _____

_____

Type of consequence _____

_____

c   Provide two reasons why the animal trainer would use a negative stimulus in the first training method.   2 marks

1. _____

_____

_____

2. _____

_____

_____

End of Question 5

Question 6 (5 marks) ⬤⬤⬜

The hippocampus and amygdala are important parts of the brain associated with our ability to remember incoming information.

a   In which part of the brain are the hippocampus and amygdala found?   1 mark

_____

b   What is the role of each of these parts in relation to memory?   2 marks

Hippocampus _____

_____

Amygdala _____

_____

c   Explain how the hippocampus and amygdala interact and work together and what this allows our memory to do.   2 marks

_____

_____

_____

_____

Question 7 (3 marks) ⬤⬤⬜

a   Explain the meaning of the term 'Country' for Aboriginal and Torres Strait Islander peoples, and how Country is used to hold ancestral knowledge.   2 marks

_____

_____

b   Explain how this use of the concept of Country is similar to method of loci.   1 mark

_____

_____

_____

End of Question 7

Question 8 (4 marks) ⚫⚫

a   Explain how our episodic and semantic memories work together to predict possible futures.   2 marks

_____

_____

_____

_____

b   Explain why some people, when predicting possible futures, cannot visualise their prediction.   2 marks

_____

_____

_____

_____

Question 9 (4 marks) ©VCAA 2018 SB Q5 (ADAPTED) ⚫⚫

a   The figure below is a hypnogram representing the sleep cycle of a healthy adult.

Sleep cycle of a healthy adult

Outline **two** differences between rapid eye movement (REM) sleep and non-rapid eye movement (NREM) sleep evident in the hypnogram above.   2 marks

_____

_____

_____

_____

Question 9 continues on page 88

b Compare how REM and NREM sleep would differ in a hypnogram of a healthy adolescent and a hypnogram of an elderly person.   2 marks

Question 10 (8 marks) ⬛⬛⬛

a Describe one difference and one similarity of circadian and ultradian rhythms.   2 marks

b Explain two likely causes of delayed sleep phase disorders.   2 marks

c How can light therapy be used to help treat circadian phase disorders such as delayed sleep onset?   2 marks

d How do blue light and natural light each help and/or harm the natural sleep–wake cycle?   2 marks

End of Question 10

Question 11 (15 marks) ⬤⬤⬤

Dr Shepherd and Dr Yang have been running a clinical trial to try to slow down the emergence of symptoms associated with Alzheimer's disease. They obtain the hospital records of patients diagnosed with Alzheimer's disease and select every fourth patient on an alphabetised list. This yields a group of participants who are displaying symptoms of the disease at various stages.

Shepherd and Yang's treatment involves injecting a new drug into patients' brains to try to improve neural communication. Each patient's treatment begins with the hospital director handing the patient's doctor a syringe marked either red or blue. The patient is not told which colour contains the treatment and which colour contains a placebo. The patients are also not told whether they have received the treatment or not.

After 1 month, the patients report on a five-point scale whether they feel their symptoms have: diminished drastically (5), diminished partially (4), not changed (3), increased partially (2) or increased drastically (1).

The results are collated and analysed by the hospital director.

a   Write an extended response outlining key features of this research that includes:

- a hypothesis for the study

- identification of the independent and dependent variables

- a summary of the sampling technique used and its appropriateness

- identification of the procedure being implemented and a discussion of why this is important.   10 marks

Question 11 continues on page 90

b   Identify an extraneous variable that may influence the results. Explain its impact and how it may be addressed in future.   3 marks

c   Discuss any ethical issues that may have been raised as important considerations for this research.   2 marks

Question 12 (4 marks) ⬤⬤⬤

Discuss the importance of a holistic approach to mental wellbeing, particularly in the context set out by the social and emotional wellbeing (SEWB) framework for Aboriginal and Torres Strait Islander peoples. Show your understanding of this model by including two different prioritised domains as outlined by the framework.

End of Question 12

Question 13 (3 marks) ⬤◯◯

When looking at risk factors that contribute to the onset of mental illness, explain one biological, one psychological and one social risk factor that can contribute to the experience of mental illness, such as a specific phobia.

Biological factor _____

_____

Psychological factor _____

_____

Social factor _____

_____

END OF PAPER

SECTION B EXTRA WORKING SPACE

STUDENT NUMBER

Letter

PSYCHOLOGY

Written examination

Reading time: 15 minutes

Writing time: 2 hours 30 minutes

QUESTION AND ANSWER BOOK

Structure of book

Section	Number of questions	Number of questions to be answered	Marks
A	50	50	50
B	12	12	70
		TOTAL	120

- Students are permitted to bring into the examination room: pens, pencils, highlighters, erasers, sharpeners and rulers.
- Students are NOT permitted to bring into the examination room: blank sheets of paper and/or correction fluid/tape.
- No calculator is allowed in this examination.

Materials supplied
- Question and answer book of 22 pages
- Additional space is available at the end of the book if you need extra space to complete an answer.

Instructions
- Write your **student number** in the space provided above on this page.
- Check that your **name** and **student number** as printed on your answer sheet for multiple-choice questions are correct, **and** sign your name in the space provided to verify this.
- Unless otherwise indicated, the diagrams in this book are not drawn to scale.
- All written responses must be in English.

Section A: Multiple-choice questions

Instructions for Section A
Answer **all** questions in pencil on the answer sheet provided for multiple-choice questions.
Choose the response that is **correct** or that **best answers** the question.
A correct answer scores 1; an incorrect answer scores 0.
Marks will **not** be deducted for incorrect answers.
No marks will be given if more than one answer is completed for any question.

Question 1 [●□□]

Which of the following nervous systems is responsible for maintaining homeostasis?

A   sympathetic nervous system

B   peripheral nervous system

C   somatic nervous system

D   parasympathetic nervous system

Question 2 [●□□]

Which of the following actions is **not** a result of somatic nervous system functioning?

A   serving a ball in tennis

B   opening a can of drink

C   playing soccer

D   breathing

Question 3 [●●□]

Which of the following describes the sequence of events in a spinal reflex?

A   Motor information travels via motor neurons to the spinal cord; it is intercepted by an interneuron, then
     a sensory neuron takes sensory information back to the site.

B   Sensory information travels via sensory neurons to a motor neuron that takes motor information back to the site.

C   Sensory information travels via sensory neurons to the spinal cord; it is intercepted by an interneuron,
     then a motor neuron takes motor information back to the site.

D   An interneuron takes information to the spinal cord; it is intercepted by a motor neuron that returns
     motor information back to the site.

Question 4 ©VCAA 2020 SA Q1 [●●●]

Which of the following correctly identifies the specialised structure and corresponding function at any given synapse?

	Structure	Function
A	pre-synaptic neuron	releases neurotransmitters from vesicles
B	synaptic gap	electrical charge transmits the neural message
C	receptor site	neurotransmitters are stored
D	post-synaptic neuron	reuptake of neurotransmitters occurs

Question 5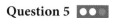

After a serious car accident, Aaliyah loses part of the temporal lobe in her brain that controls hearing. Over time, her hearing recovers. This is largely attributed to

A   a tumour.

B   practice.

C   developmental plasticity.

D   adaptive plasticity.

Question 6

Which of the following statements about glutamate is true?

A   Glutamate is an electrical impulse.

B   Glutamate is an excitatory neurotransmitter.

C   Glutamate is an inhibitory neurotransmitter.

D   Glutamate is both an inhibitory and excitatory neurotransmitter.

Question 7

Mika has just made some brownies. As she pulls them out of the oven, she can already anticipate their taste. Which neurotransmitter is giving her this sensation?

A   glutamate

B   GABA

C   serotonin

D   dopamine

Question 8

A physiological effect of prolonged arousal is _____ , while a potential psychological effect of prolonged arousal is _____ .

A   irritability; fatigue

B   dizziness; headaches

C   stomach ulcers; anxiety

D   forgetfulness; heart palpitations

Question 9

According to the General Adaptation Syndrome, upon first contact with a stressor, the body's resistance to stress falls below, then rises above, its normal level of resistance. This is due to arousal of the sympathetic nervous system.

The stage described in this explanation is known as

A   countershock.

B   exhaustion.

C   resistance.

D   alarm reaction.

Question 10 ©VCAA 2020 SA Q10 ●●●

Which one of the following is the most accurate description of the role of cortisol in the stress response, according to Selye's General Adaptation Syndrome?

A stops the immune system from functioning

B increases glucose in the bloodstream and reduces inflammation

C reactivates functions that are non-essential in a fight–flight response

D provides the initial alert about a perceived threat, through the release of adrenaline

Question 11 ●●○

Paddy generally uses deep breathing to cope with stress, but realises this technique is no longer working effectively. He feels he needs to find a new technique to cope with stress.

Paddy's realisation is known as _____ and his response is a _____ way of dealing with stress.

A burnout; negative

B an approach strategy; positive

C coping flexibility; positive

D an avoidance strategy; negative

Question 12 ●●●

According to learning theory, which of the following events characterises the first phase in classical conditioning?

A A UCS produces a UCR.

B A CS produces a CR.

C A CS produces a UCR.

D An NS produces a CR.

Use the following information to answer Questions 13–15.

The light switch in Tara's bedroom is faulty. Every time she touches the switch she gets a mild electric shock. After this has happened a few times, Tara starts to sweat when she sees a light switch anywhere in her house.

Question 13 ●●○

In this scenario, the unconditioned stimulus is

A the light switch in her room.

B the light switches around her house.

C the electric shock.

D sweating at the sight of the light switch.

Question 14 ●●○

In this scenario, the conditioned response is

A the light switch in her room.

B the light switches around her house.

C the electric shock.

D sweating at the sight of the light switch.

Question 15 ◖◖◗

Tara's family move in with a relative while their house is rewired. What is most likely to happen when Tara sees a light switch by the end of her 2-week stay at her relative's house?

A   Tara will start to sweat again.

B   Tara's conditioned response will have been extinguished.

C   Tara's unconditioned response will have been extinguished.

D   Tara will have a phobia of light switches.

Question 16 ◖◖◗

A key difference between negative and positive reinforcement is that

A   negative reinforcement removes an unpleasant stimulus, whereas positive reinforcement gives a pleasant stimulus.

B   negative reinforcement reduces the likelihood of a behaviour reoccurring, whereas positive reinforcement increases the likelihood of a behaviour reoccurring.

C   negative reinforcement removes a pleasant stimulus, whereas positive reinforcement gives a pleasant stimulus.

D   negative reinforcement increases the likelihood of a behaviour reoccurring, whereas positive reinforcement decreases the likelihood of a behaviour reoccurring.

Question 17 ◖◖◗

In classical conditioning, the learned response is _____, while in operant conditioning, the learned response is _____.

A   voluntary; reflexive

B   spontaneous; involuntary

C   reflexive; voluntary

D   involuntary; reflexive

Question 18 ◖◖◗

Aboriginal and Torres Strait Islander cultures rely on learning techniques as part of a system. Which of the following is the **least** accurate description of these techniques?

A   Learning is linear.

B   Learning involves links to the land.

C   Learning is linked to Country.

D   Learning involves visualisation.

Question 19 ◖◖◗

Long-term memory is thought to have unlimited capacity because

A   people can access long-term memories throughout their lives when given cues.

B   people can remember limitless amounts of information.

C   forgetting is a myth.

D   even the elderly can learn new things.

Question 20 ●●●

Which of the following statements about declarative memories is **true**?

A   There are two different types of declarative memories: procedural and episodic.

B   Declarative memories are also known as 'implicit memories'.

C   Declarative memories are involved with 'knowing that'.

D   Declarative memories are memories of specific facts alone.

Question 21 ●●

While getting ready for her Year 12 formal, Shien had her favourite song on repeat. Later that night, the DJ played the song and Shien was surprised to find she knew all the lyrics. Her memory of the lyrics was most likely

A   an explicit memory.

B   a photographic memory.

C   a semantic memory.

D   an implicit memory.

Question 22 ●○○

Which part of the brain is most closely associated with the memory consolidation process?

A   the reticular formation

B   the cerebral cortex

C   the hippocampus

D   the corpus callosum

Question 23 ●○○

If someone is said to suffer from aphantasia, they

A   cannot produce mental visual imagery.

B   have personality changes.

C   have short-term memory loss.

D   have long-term memory loss.

Question 24 ●●○

Which of the following statements best describes Alzheimer's disease?

A   Alzheimer's disease is a mental illness.

B   Alzheimer's disease only affects the elderly.

C   Alzheimer's disease is contagious.

D   Alzheimer's disease is a neurodegenerative disease.

Question 25 ⬤⬤⬤

Spencer is 25 and goes to bed at 11 p.m. If he sleeps continuously, at what time will he wake most naturally?

A   6 a.m.

B   6.30 a.m.

C   7 a.m.

D   7.30 a.m.

Question 26 ⬤⬤⬤

Beta waves are experienced most commonly during _____ . They are characterised by _____ frequency and _____ amplitude.

A   NWC; high; low

B   NWC; low; high

C   an ASC; high; low

D   an ASC; low; high

Question 27 ©VCAA 2020 SA Q32 ⬤⬤⬤

Sleep changes as we age.

Which one of the following statements best describes a noticeable change that occurs?

A   The time spent sleeping overall increases as we age.

B   The proportion of time spent in NREM sleep increases as we age.

C   As we progress over time from infancy to old age, the proportion of time spent in stages 1 and 2 of NREM sleep significantly decreases.

D   The proportion of time spent in REM sleep significantly decreases from infancy and then remains steady as we continue ageing.

Question 28 ⬤⬤○

India is riding in a human-powered vehicle competition and has gone without sleep for 24 hours.

Which of the following symptoms is she **least** likely to experience?

A   difficulty concentrating

B   droopy eyelids

C   hallucinations

D   mistakes on simple tasks

Question 29 ⬤⬤⬤

After several days of sleep deprivation, people often experience microsleeps. What sort of brainwaves are recorded when someone is having a microsleep?

A   alpha and theta waves

B   delta waves

C   beta and theta waves

D   sleep spindles

Use the following information to answer Questions 30–34.

Parminder compared the effects of consumption of alcohol on reaction times in people at various stages of life.

His stratified sample included participants aged 18 to 70 years. In the repeated-measures experiment, participants consumed one standard drink of alcohol at half-hourly intervals until they reached 0.10% blood alcohol concentration (BAC). Participants completed a series of computer-based tests for reaction times at BACs of 0.00%, 0.05% and 0.10%.

Additionally, once participants reached 0.10% BAC, Parminder asked all participants to write down on a lined piece of paper their immediate feelings, thoughts and memories, and to provide an estimate of how long they thought the tests ran for.

The graph below represents reaction time, in seconds, versus age, with the lines representing the trend of results for each level of BAC.

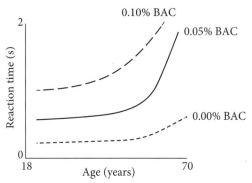

Question 30 ©VCAA 2020 SA Q33 ●●●

The graph above demonstrates that

A   an altered state of consciousness is achieved.

B   the higher the BAC, the greater the reaction time.

C   an improvement in reaction times occurs as age increases.

D   reaction times decrease significantly in both the 0.05% and 0.10% BAC conditions.

Question 31 ©VCAA 2020 SA Q34 ●●

Which of the following includes both an independent variable and a dependent variable for Parminder's study?

	Independent variable	Dependent variable
A	age	reaction time
B	reaction time	BAC
C	cognitive performance	amount of alcohol consumed
D	amount of alcohol consumed	BAC

Question 32 ©VCAA 2020 SA Q35 ●

Parminder believed when the participants were in the 0.10% BAC condition, they achieved an altered state of consciousness at the end of the study.

The most likely indication of participants being in an altered state of consciousness at the end of the study would be if they

A   reported being in control of their emotions.

B   recalled their stream of thoughts during the study.

C   remembered the conversations of passers-by outside the laboratory.

D   estimated the study going for a shorter or longer time than it did.

Question 33 ©VCAA 2020 SA Q36 ●●●

Which of the following accurately describes the two types of data Parminder was gathering during the testing period and after the last test?

A subjective and qualitative

B self-report and qualitative

C objective and quantitative

D quantitative and qualitative

Question 34 ©VCAA 2020 SA Q37 ●●

In a second study, not involving alcohol, with different participants, Parminder examined the effects on reaction time of being awake for 24 hours.

The results from Parminder's second study were likely similar to which one of the following results from his first study?

A the 0.10% BAC condition, sleep affected

B the 0.05% BAC condition, mood affected

C the 0.10% BAC condition, cognition affected

D the 0.05% BAC condition, concentration affected

Question 35 ●●●

One reason shift workers find it so hard to sleep during the day is because

A their circadian rhythm increases body temperature during the day.

B their circadian rhythm increases melatonin levels during the day.

C the noise outside is too loud.

D they are not yet tired enough.

Question 36 ●●

Which of the following is a psychological risk factor for the development of mental illness?

A low self-efficacy

B genetics

C loss of social network

D resistance to medication

Question 37 ●

Mental ill health can carry a stigma, and some people fear being labelled. What sort of risk factor is this for the development of mental illness?

A biological

B psychological

C social

D genetic

Question 38 ⬤◯◯

Which of the following factors is **least** likely to enhance someone's resilience?

A   support from family

B   insufficient sleep

C   a community network

D   adequate diet

Question 39 ⬤⬤⬤

Phobias can be treated using a variety of techniques, including systematic desensitisation. Why does systematic desensitisation have to happen over time?

A   so that an association can be formed between two responses

B   so that the phobic stimulus can be exposed in increasing intensity

C   so that the consequences for the actions can be learned

D   because patients have to be at least 18 years old, so candidates for the treatment have to wait until then

Question 40 ⬤◯◯

The social and emotional wellbeing (SEWB) framework for Aboriginal and Torres Strait Islander peoples is a _____ approach to mental wellbeing with principles that acknowledge the importance of _____ for Aboriginal and Torres Strait Islander peoples.

A   specific; culture and history

B   specific; brain and body

C   holistic; brain and body

D   holistic; culture and history

Use the following information to answer Questions 41 and 42.

Tuan migrated to Australia to undertake a university degree and lived in a share house with other students. He managed his money according to methods used in his birthplace, but he found it difficult to pay his bills on time. When his housemate asked for Tuan's rent money, Tuan's heart felt like it was pounding. Whenever he saw his housemate afterwards, Tuan would experience the same physiological sensation. When he started working, Tuan found he experienced an upset stomach and a pounding heart if someone at work suggested a different approach to his. He even started to experience these symptoms when he was in new social situations with friends. It was not long until he began to lie awake at night thinking about all of the things that were likely to go wrong the next day.

While discussing his symptoms with his doctor, the doctor suggested Tuan had a mental health problem.

Question 41 ©VCAA 2020 SA Q46   ⬤⬤⬤

Which of the following most accurately describes the type of stress Tuan was experiencing with his housemate and at work?

	With his housemate	At work
A	acculturative stress	daily pressures
B	chronic stress	distress
C	daily pressures	acculturative stress
D	distress	major stress

Question 42 ©VCAA 2020 SA Q47 ●●●

Tuan's doctor diagnosed him with anxiety. The doctor was likely to arrive at this conclusion because

A   Tuan is fearful when the outcome is not predictable.

B   Tuan's stress is ongoing and consistent with his normal behaviour.

C   Tuan's symptoms reveal that he is highly aroused and feels he cannot cope.

D   external factors, such as Tuan's housemate and his work colleagues, have contributed to his symptoms.

Question 43 ●●●

One effective treatment in combating the biological impact of anxiety is the use of benzodiazepines. Benzodiazepines are used specifically to

A   block the stress response.

B   inhibit the stress response.

C   inhibit GABA production.

D   mimic the GABA response.

Question 44 ●○○

What does the cultural determinant of self-determination mean, according to the social and emotional wellbeing (SEWB) framework?

A   an evaluation of oneself within a culture

B   values being passed down from one generation to another

C   an individual's right to shape their own life so they live according to their values

D   an individual's determination to represent their culture

Question 45 ●●○

Which statement best describes the main difference between Western and Aboriginal and Torres Strait Islander approaches to wellbeing?

A   Western culture explores the body and not the mind; Aboriginal and Torres Strait Islander cultures explore the mind and not the body.

B   Western cultures have a stronger emphasis on connection to land than Aboriginal and Torres Strait Islander cultures.

C   Aboriginal and Torres Strait Islander cultures have a stronger emphasis on connection to Country than Western cultures.

D   All of the above.

Question 46 ©VCAA 2019 SA Q50 ●●●

Xuan is a researcher who wants to gather subjective and descriptive data from people who have been diagnosed with a mental illness in order to understand what living with a mental illness is like.

Which of the following identifies the type of data Xuan is collecting, the best method for collecting this data and the best sample size?

	Type of data	Data collection method	Sample size
A	qualitative	interviews	small
B	quantitative	interviews	large
C	qualitative	questionnaire	large
D	quantitative	online questionnaire with rating scales	small

Use the following information to answer Questions 47–50.

Ravi conducted research to find out whether the coping strategy people used would affect their baseline levels of stress. Twenty participants were exposed to simulations of two different stressful scenarios. In the morning, the participants were told to use an avoidance strategy while exposed to the first simulation and, in the afternoon, they were told to use an approach strategy while exposed to the second simulation. An electroencephalograph (EEG) and electromyograph (EMG) were used to measure levels of arousal before and during the simulations. Higher levels of arousal indicated greater stress.

Readings from the EEG and EMG were quantified as stress level scores from 0 to 10. A change score was calculated by subtracting the pre-simulation stress level score from the during-simulation stress level score.

Question 47 ©VCAA 2019 SA Q10 ●●●

The dependent variable was operationalised as the

A coping flexibility of the strategy.

B stress level score calculated from EEG and EMG measurements.

C levels of arousal during the simulations as measured by the EEG and EMG.

D change score calculated as the difference between pre-simulation and during-simulation stress level scores.

Question 48 ©VCAA 2019 SA Q11 ●●●

Which one of the following was a confounding variable in Ravi's research?

A using the same participants in both conditions, as there may be practice effects

B using only 20 participants, as this does not allow for generalisation of the results

C telling participants to use a particular coping strategy, as this may bias participants

D using only one strategy in each condition, as this does not allow for coping flexibility

Question 49 ©VCAA 2019 SA Q12 ●●

Physiological measurements of arousal were made before and during the simulations. This allowed Ravi to

A detect changes in levels of arousal.

B compare the differences in strategies.

C check the reliability of the measurements.

D eliminate participants whose levels of stress have not changed.

Question 50 ©VCAA 2019 SA Q13 ●●

Ravi hypothesised that an avoidance strategy would be more likely to result in a bigger increase in levels of stress. Which of the following supports Ravi's hypothesis?

	Mean change score for avoidance strategy	Mean change score for approach strategy
A	+2.2	+6.2
B	+6.2	+1.5
C	−2.2	−3.2
D	+2.2	−1.5

Section B

> **Instructions for Section B**
> Answer all questions in the spaces provided.

Question 1 (5 marks)

a Many mental and physical health problems occur as a result of too much or insufficient amounts of particular neurotransmitters. Explain what neurotransmitters are, then describe the likely effects on the body of too much glutamate, and the likely effects on the body of not enough glutamate.   3 marks

b Using examples to illustrate your answer, explain how the neurotransmitter dopamine can lead to both helpful and harmful behaviours.   2 marks

Question 2 (15 marks) ©VCAA 2020 SB Q2

For his Psychology practical investigation, Peter decided to examine whether information about the positive effects of stress could affect stress levels during public speaking. His participants were 20 newly appointed volunteer leaders at a local organisation. Peter randomly assigned participants to one of two groups.

Participants in the experimental group were told they would be recorded giving a five-minute speech one week later and were shown a brief presentation outlining how stress can improve performance. They were told not to discuss this presentation with participants in the other group.

Participants in the control group were only told they would be giving a five-minute speech one week later and that their speech would be recorded.

The heart rate of participants in both groups was measured prior to them being told about the five-minute speech, in order to provide a baseline, and one week later, immediately after giving the five-minute speech. Means for the two measurements for each group were calculated. The results are shown in the graph.

Question 2 continues on page 106

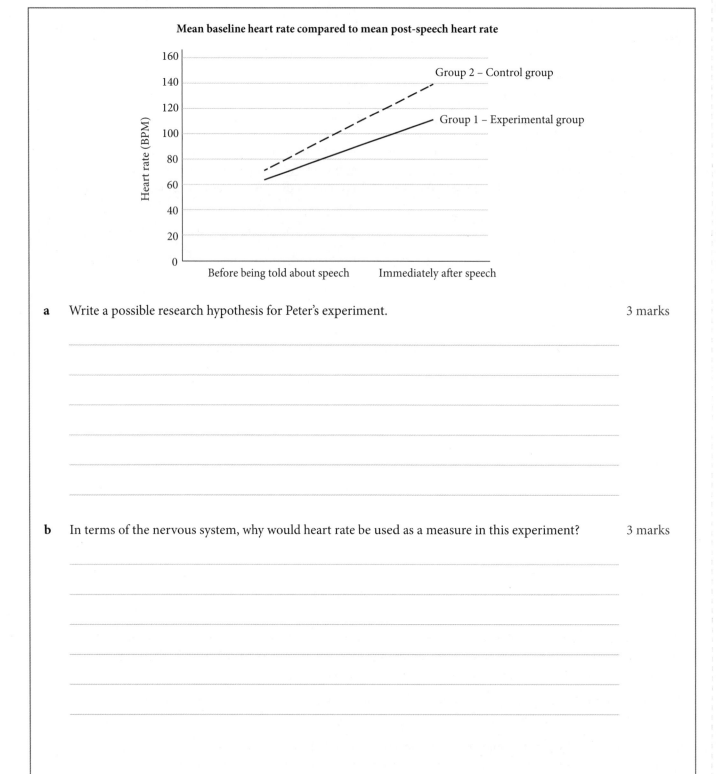

Mean baseline heart rate compared to mean post-speech heart rate

a   Write a possible research hypothesis for Peter's experiment.   3 marks

_____

_____

_____

_____

_____

_____

b   In terms of the nervous system, why would heart rate be used as a measure in this experiment?   3 marks

_____

_____

_____

_____

_____

Question 2 continues on page 107

c Using Lazarus and Folkman's Transactional Model of Stress and Coping, identify the most likely secondary appraisal made by participants in the experimental group and by participants in the control group. Justify your response with reference to the graph on page 106 and the independent variable.   6 marks

Group 1 – Experimental group

Secondary appraisal _____

_____

Justification _____

_____

_____

_____

_____

Group 2 – Control group

Secondary appraisal _____

_____

Justification _____

_____

_____

_____

_____

d At the conclusion of his practical investigation, Peter realises that, entirely by chance, seven participants allocated to the experimental group were members of the region's high-performing basketball team whereas the control group contained only one participant who was an athlete.

Explain the problem created by this uneven allocation of athletes between groups, identifying the type of variable involved.   3 marks

_____

_____

_____

_____

_____

End of Question 2

Question 3 (6 marks)

a Penny has been experiencing high levels of stress following the summer holiday break. She consults her doctor, whose recommendations include clinical investigation of her gut microbiota. Explain what gut microbiota may have to do with her feelings of stress. 2 marks

b Penny's doctor suggests that she try different strategies to relieve her stress while they are investigating her gut microbiota. Discuss the main distinction between approach and avoidance strategies for dealing with stress, and comment on their effectiveness as stress-relieving techniques. 4 marks

Question 4 (7 marks) ©VCAA 2020 SB Q4

Advertisers often use learning principles when promoting products. An advertisement for a new soft drink features people having a good time while consuming the product. This is intended to make potential customers experience positive emotions when thinking about the soft drink.

a What type of conditioning is used to generate positive emotions towards the new soft drink? Give two reasons to justify your response. 3 marks

Type of conditioning

Reason 1

Reason 2

Question 4 continues on page 109

b　The next advertising campaign for the soft drink used a celebrity rather than an unknown person.

Identify two different processes involved in observational learning that demonstrate the advantage of using a celebrity to advertise the soft drink. Justify your response for each process.　　4 marks

Process 1 _____

Justification _____

_____

_____

_____

Process 2 _____

Justification _____

_____

_____

_____

Question 5 (2 marks) ⬤⬤▮

a　What is one difference between iconic and echoic memory?　　1 mark

_____

_____

b　What is one similarity between iconic and echoic memory?　　1 mark

_____

_____

Question 6 (4 marks) ⬤⬤⬤

Describe one mnemonic from written cultures and one mnemonic from oral cultures, clearly outlining an advantage of each.

_____

_____

_____

_____

_____

_____

_____

End of Question 6

Question 7 (3 marks)

Discuss three differences between the experience of REM and NREM sleep.

_____

_____

_____

_____

_____

Question 8 (3 marks)

a What does an EEG measure?   1 mark

_____

_____

b What does an EOG measure?   1 mark

_____

_____

c Label on the diagram below where the electrodes for EEGs, EOGs and EMGs are typically placed.   1 mark

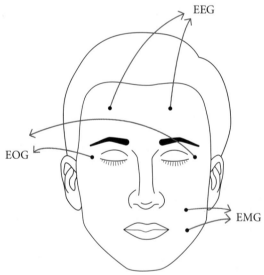

End of Question 8

Question 9 (3 marks) ▮▮▯

a Gemma recently gained her full driver's licence, and her mother is worried about the prospect of her drinking and driving. Her mother makes her sit down and write three reasons why drink-driving is dangerous.

Using your knowledge of the characteristics of an altered state of consciousness, discuss two potential effects on Gemma's driving ability if she were to drink alcohol.   2 marks

_____

_____

_____

_____

_____

b Compare the negative effects associated with drink-driving with those experienced after one night's sleep deprivation.   1 mark

_____

_____

Question 10 (6 marks) ▮▮▯

Many educators are concerned about the impact of electronic screens on their students' sleep–wake cycle. Explain why this is the case, and provide some advice that educators could give to students to improve their sleep–wake cycle. Why are educators worried about students getting enough sleep?

_____

_____

_____

_____

_____

_____

_____

_____

End of Question 10

Question 11 (10 marks)   ©VCAA 2018 SB Q4 ●●●

EXERCISE MAY BE THE KEY TO DELAYING THE PROGRESSION OF ALZHEIMER'S DISEASE
by Helena Graviana

A research team led by the University of Pittsburgh and the University of Illinois investigated how regular exercise may help protect the brain from the normal age-related decline of brain structures that are important for memory and learning.

The researchers randomly allocated 120 healthy older adult volunteers to either a moderate-intensity aerobic exercise group (brisk walking) or a low-intensity exercise group (stretching). Each group was led by an exercise instructor for 40 minutes, three times a week for one year.

Brain images (using magnetic resonance imaging, or MRI) were taken of each participant to measure the volume of the hippocampal structures at the beginning of the study, after six months and after one year. The groups differed slightly in mean hippocampal volume at the beginning of the study but, after one year, the moderate-intensity aerobic exercise group demonstrated a 2% increase in volume, whereas the low-intensity exercise group demonstrated a 1.4% decrease.

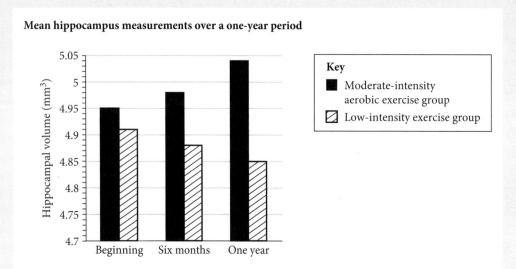

Mean hippocampus measurements over a one-year period

The graph above shows the mean hippocampal volume (in cubic millimetres) for the left and right hemispheres for the moderate-intensity aerobic exercise group and the low-intensity exercise group over a one-year period.

The results provide evidence that regular moderate-intensity aerobic exercise can reverse the normal age-related decline in hippocampal volume, and that this could protect against the development and progression of Alzheimer's disease.

Data: adapted from KI Erickson, MW Voss, RS Prakash, et al., 'Exercise training increases size of hippocampus and improves memory', in *Proceedings of the National Academy of Sciences of the United States of America*, vol. 108, no. 7, 15 February 2011, pp. 3017–3022

a   Write a research hypothesis for this investigation.   3 marks

_____

_____

_____

_____

_____

Question 11 continues on page 113

b The researchers were unable to generalise the results from this investigation to people with Alzheimer's disease.

Why? 1 mark

c Explain why the effects of exercise on hippocampal volume may be of interest to researchers investigating Alzheimer's disease. 3 marks

d Researchers decided to conduct a similar investigation using people diagnosed with Alzheimer's disease.

Describe the informed consent procedures that should be used by researchers investigating people diagnosed with Alzheimer's disease. 3 marks

Question 12 (6 marks)

The biopsychosocial framework consists of overlapping biological, psychological and social factors that contribute to the experience of mental illness. Name two factors for each of the three categories that can contribute to the incidence of a person's specific phobia.

Biological

Psychological

Social

END OF PAPER

SECTION B EXTRA WORKING SPACE

ANSWERS

Test 1: Nervous system functioning

Multiple-choice answers

Question 1

C somatic

The other branch of the PNS is the autonomic nervous system, which has two subdivisions, as pictured. **A** is incorrect, because the central nervous system is a separate nervous system to the peripheral nervous system and therefore not a component of it. **B** is incorrect because the autonomic system has two subdivisions and is represented by the letter B in the picture. **D** is incorrect because the sympathetic nervous system is a subdivision of the autonomic nervous system and is represented by the letters C or D in the picture.

Question 2

D sympathetic

D could have been either the sympathetic or parasympathetic nervous system: there is no difference. These are the subdivisions of the autonomic nervous system. **A** is incorrect because the central nervous system is a separate nervous system to the peripheral nervous system and therefore not a component of it. **B** is incorrect because letters C and D represent subdivisions of the autonomic nervous system. **C** is incorrect as the somatic nervous system is represented by the letter A in the picture, as it has no further subdivisions.

Question 3

D The autonomic nervous system is responsible for regulating physiological responses that do not require conscious control.

The autonomic nervous system functions without conscious control. It does have two subdivisions, but not the two listed in the question, and it is not responsible for conscious decisions such as those to do with integration of information. **A** is incorrect because the autonomic nervous system is responsible for involuntary responses and regulation of non-skeletal muscles. The somatic nervous system is responsible for the voluntary control of skeletal muscles. **B** is incorrect because the somatic nervous system is not a subdivision of the autonomic nervous system. The two subdivisions of the autonomic nervous system are the sympathetic and parasympathetic nervous systems. **C** is incorrect because the nervous system responsible for integrating sensory and motor information is the central nervous system.

Question 4

B decreased saliva, dilated pupils

Several bodily functions indicate a heightened state of arousal. Functions necessary for survival, such as sight, increase (dilated pupils allow more light in), while those less important, such as saliva production, decrease. **A** is incorrect because there would be an increase in both heart rate and lung capacity/breathing rate in a heightened state of arousal: both changes allow for more oxygen-rich blood to be diverted to the muscles to increase the chances of survival. **C** is incorrect because while lung capacity will increase, saliva production will decrease. This is because production of saliva is not required for survival in times of stress and threat. **D** is incorrect because pupils will dilate to allow more light in, not constrict. Heart rate will increase.

Question 5

B digestion increases

The opposite must now occur to arousal: while heart rate and so on comes down, those functions suppressed, such as digestion, are stimulated. **A** is incorrect because once the body calms down after a threat, heart rate will decrease, not increase. Increased heart rate occurs when we're faced with a threat and the sympathetic nervous system is dominant. **C** is incorrect because bile is a digestive fluid, and when the parasympathetic nervous system is dominant, digestion returns to normal levels. Therefore, the release of bile will be stimulated. **D** is incorrect because saliva production increases when the parasympathetic nervous system is dominant. This is because the body no longer has to shut down processes unnecessary for survival, such as saliva production.

Question 6

D spinal reflex

The spinal reflex is the only response not coordinated by the brain. It is initiated by interneurons in the spinal cord. **A** is incorrect because breathing is regulated by the medulla in the brain. **B** is incorrect because reading involves several parts of the brain. For example, the occipital lobes, which organise the incoming visual stimuli. **C** is incorrect because Broca's area in the left frontal lobe of the brain is responsible for speech production.

Question 7

A the adaptive nature of the human nervous system

Only 17% of students who sat the VCE exam answered this question correctly. A spinal reflex is adaptive in nature, as it enhances chances of survival. For example, quickly removing oneself from harm, as described in the question. **B** is incorrect because the spinal cord does not make decisions: the brain is responsible for decision-making. While the withdrawal response above is a spinal reflex, the wording of this answer makes this option incorrect. **C** is incorrect because a reflex is an unconscious response (i.e. done without awareness), not a conscious one. **D** is incorrect because the autonomic nervous system regulates involuntary responses. Also, the brain is not involved in the spinal reflex (the withdrawal response described in the question).

Question 8

B electrochemical energy.

Electrical energy is used within neurons, but chemical energy is used between them. Therefore, electrochemical energy is needed for neurons to communicate. **A** and **C** are incorrect because electrical energy is only part of neural transmission and communication. **D** is incorrect because neural energy is a vague term and not specific enough for VCE Psychology.

Question 9

A a gap between two neurons

The gap between two neurons is known as a 'synaptic gap', and this is the point of communication between two neurons (the presynaptic and postsynaptic neurons). **B** is incorrect because neurotransmitter is the chemical that is released from the presynaptic neuron into the synapse and that crosses the synapse to the postsynaptic neuron. **C** is incorrect because an action potential is the neural impulse, not the gap between two neurons. **D** is incorrect because resting potential explains how and when a neuron is waiting for activation.

Question 10

D to stimulate a response

As the name suggests, an excitatory neurotransmitter stimulates or creates the initiation of a response. **A** is incorrect because an excitatory neurotransmitter is likely to elevate mood. **B** is incorrect because an excitatory neurotransmitter makes it more likely for the postsynaptic neuron to fire, and therefore initiate a response. **C** is incorrect because calming of the CNS is more likely to occur due to an inhibitory neurotransmitter such as GABA.

Question 11

B neuroplasticity.

Neuroplasticity is the brain's ability to change and adapt and is shown through physiological changes. These changes can include changes to dendrites, to axon terminals and within the neurotransmitter production. **A** is incorrect because the changing of neurons and connections occurs throughout the brain, not just in the hippocampus. **C** is incorrect because synaptic formation involves changes to the connections between neurons, in particular, the creation of new connections between neurons. **D** is incorrect because neurotransmission refers to how neurons communicate with each other, not how neurons change due to experience.

Question 12

B synapse

Memories are stored in the synapses between neurons, and in the wiring they have to one another. **A** is incorrect because the dendrites are responsible for receiving neurotransmitter from the synapse. **C** is incorrect because the axon is responsible for transmitting the neural impulse from the cell body to the axon terminals. **D** is incorrect because the myelin sheath is the protective layer that coats the axon and allows for the quick and efficient transmission of the neural impulse.

Question 13

C the presynaptic neuron and postsynaptic neuron are firing at the same time.

The well-known Hebb's rule, 'neurons that fire together wire together', is how neuroplasticity occurs through long-term potentiation and the strengthening of a neural connection. All other options are incorrect because the presynaptic and postsynaptic neurons need to be firing together for neural pathways to be strengthened or for long-term potentiation to occur.

Question 14

A There is a change in neurons when learning takes place.

Learning is a physical process that creates relatively permanent change in behaviour and in the neurons that are shaped by the experience. **B** is incorrect because at the most fundamental level, learning is the result of changes to neural connections; therefore, neurons are affected by learning. For example, LTP and LTD. **C** is incorrect because anything temporary would not necessarily be considered a learned behaviour, as learning is a relatively permanent change. **D** is incorrect because the number of neural connections will increase as learning takes place.

Question 15

C transmit sensory messages to the brain.

A neuron has many roles, so answering this question is a process of elimination, and the answer does not reflect the entire picture. Neurons don't travel, they transmit messages, counting out **A** and **B**. **D** is incorrect, because sensory or afferent go towards the brain, motor or efferent go away from it.

Question 16

C central nervous system.

Although a spinal reflex is indeed an automatically occurring response, it is initiated in the spinal cord. The spinal cord is part of the central nervous system. **A** is incorrect, because the peripheral nervous system is responsible for transmitting information to and from the central nervous system, not initiating the survival mechanism of the spinal reflex. **B** is incorrect because while the autonomic nervous system is responsible for unconscious and involuntary responses, these responses are of non-skeletal muscles, organs and glands, therefore not initiating motor movement of skeletal muscles. **D** is incorrect because the sympathetic nervous system becomes activated when someone perceives themself to be in danger.

Question 17

A neurotransmitter.

GABA is an inhibitory neurotransmitter. This means that it calms neural activity. **B** is incorrect because GABA is a neurotransmitter, meaning it is released from the presynaptic neuron into the synapse to affect the activity of the postsynaptic neuron. **C** is incorrect because GABA is a neurotransmitter produced by neurons, not a neuron itself. Neurons include structures such as dendrites, axons and axon terminals. **D** is incorrect because GABA is not a structure of the brain, but a neurotransmitter found in the brain.

Question 18

B decrease a feeling of anxiety.

As GABA is an inhibitory neurotransmitter, it decreases a response. GABA is known for its role in stress and anxiety, calming the body's response to feared stimuli. **A** is incorrect because the neurotransmitter generally involved in pain is the excitatory neurotransmitter glutamate. **C** is incorrect because it is generally the prevention of the release of neurotransmitter that reduces the sensation of pain. **D** is incorrect because GABA is an inhibitory neurotransmitter, not an excitatory one.

Question 19

A learning and memory.

Being an excitatory neurotransmitter, glutamate stimulates a response. In this case, it is the ability to learn information by creating the potential for neurons to create memories. **B** and **D** are incorrect because the neurotransmitter involved in motor movement, reward and motivation is dopamine. **C** is incorrect because long-term depression is the long-lasting decrease in the strength of a synaptic connection; therefore, the neurons have not been firing together. Glutamate makes it more likely for neurons to fire together.

Question 20

A memory

Memory formation is the responsibility of glutamate, while reward, pleasure and motivation are all functions of dopamine, which can lead to motivating behaviours and addictive behaviours. **B** is incorrect because dopamine is commonly associated with reward, therefore strengthening a reward neural pathway in the brain. **C** is incorrect because dopamine is commonly associated with pleasure. It helps us feel pleasure as part of the reward systems set up in the brain. **D** is incorrect because dopamine is commonly associated with motivation. It is believed to be involved in initiation and perseverance of tasks.

Short-answer sample responses

Question 1

The somatic nervous system is responsible for the voluntary control of skeletal muscles. (1 mark) The somatic nervous system is responsible for sending sensory information from the sensory receptors to the brain for processing motor information from the brain. (1 mark)

Question 2 ©VCAA 2020 EXAM REPORT SB Q1

Difference 1: The sympathetic nervous system can bring about change in visceral muscles, organs and glands (1 mark), whereas the spinal reflex brings about change in (skeletal) muscles. (1 mark)

Difference 2: The spinal reflex is controlled by the central/somatic nervous system, (1 mark) whereas the sympathetic nervous system is controlled by the autonomic nervous system. (1 mark)

Question 3

The presynaptic neuron releases neurotransmitters from the terminal button or synaptic knob. (1 mark) The neurotransmitter is a chemical that passes across the synapse. (1 mark) The postsynaptic neuron receives the information via the dendrite. (1 mark)

Question 4

a LTP is the strengthening of neural connections to enhance the effectiveness of a response. (1 mark)

b More axon terminals or branches are grown. (1 mark) Dendrites grow more dendritic spines or become 'bushier'. (1 mark)

Question 5

a Developmental plasticity is the brain's ability to create new neurons that make new connections. (1 mark) In developmental plasticity, neural connections strengthen or are pruned to enable increased developmental opportunities. (1 mark)

b Adaptive plasticity occurs when the brain responds to damage within the brain. (1 mark) The neurons adapt and re-route to create new neural potential for damaged areas. (1 mark)

c Adaptive plasticity has increased capability at the same time as developmental plasticity is enhanced. These times are early in life, the first few years of life and in early adolescence, from about 12–15 years. (1 mark)

Question 6

The neurotransmitter dopamine is associated with reward, motivation and pleasure. (1 mark) When someone gets a reward – even a small one – such as in gambling, dopamine is released into the brain. (1 mark) This feeling is addictive: you want the sensation of pleasure again, and so you are motivated to repeat the behaviour. (1 mark)

ANSWERS – TEST 1

Test 2: Stress as an example of a psychobiological process

Multiple-choice answers

Question 1

D cortisol

Cortisol is often referred to as the stress hormone, as it is secreted to help the body fight a stressor. However, it can leave the body susceptible to illness while it is diverted from defending the immune system. **A** is incorrect because GABA is an inhibitory neurotransmitter that is involved in calming the central nervous system. **B** is incorrect because glutamate is an excitatory neurotransmitter that is involved in memory and learning. **C** is incorrect because dopamine is a neurotransmitter that is involved in motor control, reward and motivation.

Question 2

B chronic stress.

Chronic stress is stress that is ongoing. Stress that is intense in nature and sudden is known as acute. Eustress is stress that is helpful for preparing the body for vigorous activity. **A** is incorrect because eustress is a type of stress that involves a positive psychological response, such as enthusiasm, and it is generally more short-term in nature. **C** is incorrect because acute stress involves rapid onset of stress that only lasts a short time. **D** is incorrect because a stressor is any stimulus that produces stress. This may be internal or external in nature.

Question 3

D eustress.

Eustress is stress that is helpful for preparing the body for vigorous activity. The opposite of this is distress, which creates debilitating effects. **A** is incorrect because an anxiety disorder is a diagnosable mental disorder. A person with the disorder has a range of symptoms that affect their daily functioning for a certain period. **B** is incorrect because distress involves negative psychological responses such as fear. Indra reports feeling 'in the zone' and ready to go: these are positive psychological responses. **C** is incorrect because chronic stress is ongoing: Indra would only experience these symptoms for a short time.

Question 4

D when they can't choose between fight or flight

The body naturally chooses between fight or flight; however, when the chance of survival using either of these options is not clear, a person may freeze while considering their options. Another advantage of the freeze response in terms of survival is that an inactive person is less likely to be detected. **A** is incorrect because when a person experiences stress their body quickly produces extra energy needed for survival. **B** is incorrect because when a person experiences the freeze response, their body often acts as if injured (e.g. their blood pressure drops). However, the freeze response does not happen *because* they are injured. **C** is incorrect because when stressed, a person is going to experience the fight–flight–freeze response, not necessarily the freeze response only.

Question 5

A acute

The fight–flight–freeze response is an initial reaction to a stressor. Typically, this stressor is immediate and unexpected. **B** is incorrect because chronic stress is prolonged. The body cannot maintain the high levels of arousal associated with the fight–flight–freeze response for long periods. **C** is incorrect because the fight–flight–freeze response is the immediate changes to the body when confronted with a stressor. Eustress involves a psychological response and therefore is not as immediate: it requires some interpretation of a situation. **D** is incorrect because while the fight–flight–freeze response occurs in emergency situations, the type of stress is not referred to as 'emergency stress'.

Question 6

C exhaustion stage; the General Adaptation Syndrome

While Miss Bell has coped for a while, her stress is prolonged, and she has become so unwell with the flu that she has been hospitalised, so she is most likely in the exhaustion stage of the General Adaptation Syndrome. **A** is incorrect because during the resistance stage of the General Adaptation Syndrome, a person is still able to cope with the stressor. They may experience illness, but it will be less severe. **B** is incorrect because the resistance stage is part of the General Adaptation Syndrome, not the fight–flight–freeze response. **D** is incorrect because while Miss Bell has entered the exhaustion stage, this is part of the General Adaptation Syndrome, not the fight–flight–freeze response.

Question 7

D drops below normal level, then rises above.

Resistance to stress drops below normal levels in the shock substage. The body acts as though it is injured and protects vital organs and functioning. However, the body quickly rebounds in countershock and the body's ability to deal with the stressor rises above normal levels. **A** is incorrect because in the alarm reaction stage the body's ability to deal with the stressor drops first in shock, before quickly rebounding and rising above normal levels in countershock. **B** is incorrect because this option ignores the first part of the alarm reaction stage, which is shock. In shock, the body's ability to resist the stressor drops below normal levels. **C** is incorrect because this option ignores the countershock component of the alarm reaction stage. During countershock, the body's ability to resist the stressor rises above normal levels.

Question 8

A increased immunity

In the exhaustion stage, the body's resources have been depleted, and the body is therefore more susceptible to illness and disease. **B**, **C** and **D** are all possible reactions during the exhaustion stage of stress and are therefore incorrect, meaning the answer is **A**. (Colds and flus are typical of the resistance stage of the General Adaptation Syndrome. Stomach ulcers and high blood pressure could also be seen in the exhaustion stage.)

Question 9

B resistance; Continued cortisol release weakens Jamie's immune system, resulting in his body being unable to fight the cold.

The difficulty in this question is deciding between **B** and **C**. Mild symptoms are expected in the resistance stage and a cold is generally a mild illness. The question gives no further details about Jamie's ability to cope or his resources being depleted, so **B** is the best answer. **A** is incorrect because shock is experienced when Jamie first comes into contact with the stressor, not when Jamie has been experiencing constant stress. Also, shock is not a stage of the General Adaptation Syndrome: it is a component of the alarm reaction stage. **C** is incorrect, because if Jamie were in the exhaustion stage, we would expect to see a more serious illness than a cold. **D** is incorrect because during the resistance stage, cortisol would be released into Jamie's bloodstream, making Jamie more susceptible to illness. Adrenaline is released in the alarm reaction stage to prepare the body for action.

Question 10

B she felt excited about progressing to the finals if her team won.

Eustress is the good stress that gets you 'in the zone' and gives you the resources you need. **B** is the only positive option that would enhance Masako's performance. **A** is incorrect because eustress involves positive psychological responses to stress. Doubting your abilities is not a positive response. **C** is incorrect because eustress involves positive psychological responses to stress. This can include feeling more focused on the task at hand. Therefore, Masako's inability to concentrate is not a sign of eustress. **D** is incorrect because eustress involves positive psychological responses to stress: feeling nervous is not a positive response. Rather, it suggests Masako is experiencing distress.

Question 11

B non-dominant; dominant; increased blood pressure

The body's nervous systems work together, and so describing them as dominant and non-dominant is more correct. We certainly have sympathetic dominance in this scenario, meaning the parasympathetic nervous system is non-dominant. One physiological response at this time is increased heart rate and hence increased blood pressure. **A** is incorrect because when describing the functioning of the parasympathetic and sympathetic nervous system, it is more correct to refer to them as dominant or non-dominant. They are never completely active or inactive. **C** is incorrect because the terms inactive and dominant are not uniform. The parasympathetic nervous system is never completely inactive, but at times it may not be dominant. **D** is incorrect because when describing the functioning of the parasympathetic and sympathetic nervous system, it is more correct to refer to them as dominant or non-dominant. They are never completely active or inactive. Further, one physiological response at this time would be dilated (not constricted) pupils, which would allow Masako to see more clearly.

Question 12

C of the situation as good practice for the finals.

Primary appraisal involves Masako's assessment of whether this is stressful or not, and whether it presents harm, loss, threat or challenge. This scenario was a great challenge for the task to come. **A** is incorrect because primary appraisal involves assessing whether a situation is stressful or not. The crowd cheering her on could contribute to eustress, but this is not the situation she is appraising. Rather, she is appraising the situation of going out to bat. **B** is incorrect because the tips given to her by her coach are strategies and advice. She is appraising the situation of going out to bat. **D** is incorrect because Masako thinking about how she practised is an example of strategies. She is appraising the situation of going out to bat.

Question 13

C whether it is a stressful situation or not.

The primary appraisal stage is the first stage of the model, in which a person assesses whether a situation is stressful or not. If perceived as such, they also decide whether there is harm−loss, threat or challenge. After this, they move to the secondary appraisal stage, in which they consider available resources. **A** is incorrect because deciding whether they have appropriate resources to cope is part of secondary appraisal. This is more conscious than primary appraisal. **B** and **D** are incorrect because deciding who to turn to is part of secondary appraisal, in which a person decides if they have the resources and strategies to cope. This is more conscious than primary appraisal.

Question 14

B during secondary appraisal

The primary appraisal stage is the first stage of the model, in which a person assesses whether a situation is stressful or not. After this, they move to the secondary appraisal stage, in which they consider available resources. **A** is incorrect because primary appraisal is the first stage, in which a person assesses whether a situation is stressful or not. If perceived as such, they also decide whether there is a harm−loss, threat or challenge. **C** is incorrect because the resistance stage is part of the General Adaptation Syndrome, which is a biological model. **D** is incorrect because the exhaustion stage is part of the General Adaptation Syndrome, which is a biological model.

Question 15

D challenge

The primary appraisal stage is the first stage of the model, in which a person assesses whether a situation is stressful or not. If perceived as such, they also decide whether there is a threat, harm, loss or a challenge in this threat. Challenge is the most positive of these stressful situations, as there is potential for personal growth. **A** and **B** are incorrect because a primary appraisal of harm or loss means that some form of damage has already occurred, which is more closely associated with distress. **C** is incorrect because primary appraisal of a threat means that some form of damage or loss may occur, which is more closely associated with distress.

Question 16

C there are too many demands for available resources.

According to Lazarus and Folkman, it is not the stressful situation or demands that determine the degree of stress, but the resources available for coping. Even very stressful situations or demands can be managed given adequate resources. Stress is greatest when the demands outweigh the available resources. **A** is incorrect because even very stressful situations or demands can be managed given adequate resources. **B** is incorrect because if there are no demands, it is likely a person will not experience stress. **D** is incorrect because if the available resources outweigh the demand, a person is usually able to cope.

Question 17

B gut–brain axis

Microbiota live in the gut and influence not just physical health but psychological health as well. **A** is incorrect because the HPA axis refers to the way the body continues to release cortisol during ongoing stress. **C** is incorrect because the spinal reflex is a survival reflex initiated by the spinal cord that involves the motor neurons, sensory neurons and interneurons. **D** is incorrect because the General Adaptation Syndrome is a biological model to explain the impact of stress on physiological functioning.

Question 18

A living micro-organisms

Microbiota are living bacteria or organisms that live in the gut and can influence not just physical health but also psychological health. **B** is incorrect because neurons are cells that are the fundamental building blocks of the brain and nervous systems. **C** is incorrect because neurotransmitters are chemical messages that allow neurons to communicate with each other. **D** is incorrect because stress cells are not a type of cell found in the body.

Question 19

B problem-focused coping.

Problem-focused coping involves looking for solutions and executing them. It involves making a step towards seeking a resolution. **A** is incorrect because emotional forecasting does not deal directly with a stressor. Nor is it about making a plan to resolve a problem. **C** is incorrect because emotion-focused coping involves dealing with and reducing the negative emotions associated with a stressor. It does not involve dealing directly with the stressor itself. **D** is incorrect because eustress is a type of stress involving positive psychological responses to a stressor – it is not a type of coping strategy.

Question 20

B It utilises unused adrenaline.

As well as delivering mental health benefits and benefits to do with routine, exercise uses unused adrenaline. Adrenaline is produced during times of stress, and if not utilised by the body it can lead to chronic health conditions such as stomach ulcers. **A** is incorrect because exercise will increase the amount of endorphins released. **C** is incorrect because during exercise the body is able to use up cortisol, therefore reducing the amounts of cortisol. **D** is incorrect because exhaustion is a stage associated with the General Adaptation Syndrome, whereby a person is no longer able to resist a stressor as the body is depleted of all resources.

Short-answer sample responses

Question 1

2 marks are given for any two appropriate physiological responses, including heart palpitations, colds/flu, rashes, dizziness and so on.

2 marks are given for any two appropriate psychological responses, including those that are behavioural (changes to sleep or eating patterns), emotional (mood swings and irritability) or cognitive (thoughts of hopelessness and irrational thinking).

For 1 mark, explain that physiological responses are those that influence the body, whereas psychological responses are those regarding behaviour, emotion or cognitions.

Question 2

a Increases: Responses could include: increase in heart rate, expanded lungs for respiration, dilate pupils and so on.

   Decreases: Responses could include: decrease in saliva production, digestion inhibited, bladder constricts and so on. (2 marks)

b The body uses its resources in ways that are most advantageous for survival. While maximising the use of these resources, it conserves the energy use of other resources to help optimise functioning overall. (1 mark)

Question 3

a Primary appraisal: Ruby may evaluate whether the course is creating stress or not. She may decide that the traffic and isolation are indeed stressful events.

   Secondary appraisal: Ruby may look at the supports and resources she has in place. Secondary appraisal could involve, for example, Ruby realising that her friends are an asset: she could study with them and even stay overnight at their houses to avoid driving in traffic and to ease her stress and feelings of isolation. (2 marks)

b The Transactional Model of Stress and Coping is a psychological-based theory of stress. (1 mark) This means that people may view stressors and resources differently. For example, another student in Ruby's class may enjoy the 'alone time' of independent study and of driving alone, (1 mark) whereas Ruby finds this difficult. The Transactional Model of Stress and Coping is centred around the fact that people will view situations uniquely. (1 mark)

Question 4

The gut–brain axis is a bi-directional communication link between the central and enteric nervous systems. (1 mark) Good intestinal health can lead to improved cognitive and emotional functioning of the brain, while bad intestinal health can have a negative effect. (1 mark) The gut is sometimes referred to as a 'second brain', as it communicates with our brain to work on the body's overall health. (1 mark)

Question 5

Annabelle may use an approach strategy, such as closing the café for a day to regroup with some trusted employees and work out the next steps for the business (award 1 mark for any strategy that takes action). Approach strategies are effective for coping – not just to improve mindset, but also to change the situation. (1 mark)

Annabelle could use an avoidance strategy, such as closing the shop and trying not to think about it. (1 mark) While avoidance strategies can provide a temporary break from the problem, they are not effective for moving forward or for coping in the longer term. (1 mark)

Test 3: Models to explain learning

Multiple-choice answers

Question 1

C a relatively permanent change in behaviour that occurs as a result of experience.

For learning to occur, there must be a change in behaviour. Learning can take place in many different ways. **A** is incorrect because an information-processing system is more likely referring to the process of memory rather than learning. **B** is incorrect because learning involves a relatively permanent change, not a brief change. **D** is incorrect because learning can occur unintentionally. For example, the role of the learner in classical conditioning is passive.

Question 2

C unconditioned stimulus and neutral stimulus.

In classical conditioning, the repeated association between the unconditioned stimulus (which produces the unconditioned response) and the neutral stimulus is what causes the conditioning. Eventually, the neutral stimulus becomes the conditioned stimulus and produces the conditioned response. **A**, **B** and **D** are incorrect because classical conditioning involves the repeated association of two stimuli – the unconditioned stimulus and the neutral stimulus – not the repeated association of a stimulus and response.

Question 3

A yellow balloons.

Yellow balloons were neutral before the incident. Now they are paired with the loud popping noise, they have become the conditioned stimulus. **B** is incorrect because crying from the loud noise is the unconditioned response. **C** is incorrect because crying from the yellow balloons is the conditioned response. **D** is incorrect because the balloons popping is the unconditioned stimulus.

Question 4

B crying from the loud noise.

The unconditioned response is that which occurred without conditioning: it occurred naturally. In this scenario, the unconditioned response was Kim crying at the loud noise. **A** is incorrect because the yellow balloons are the neutral stimulus, which go on to become the conditioned stimulus. **C** is incorrect because crying at the yellow balloons is the conditioned response. **D** is incorrect because the balloons popping is the unconditioned stimulus, which leads to the unconditioned response of crying from the loud noise.

Question 5

B the meat powder.

The meat powder prompts a naturally occurring unconditioned response (salivation) when it is presented. Therefore, it is an unconditioned stimulus. **A** is incorrect because Pavlov is the experimenter. An unconditioned stimulus needs to produce a naturally occurring unconditioned response: Pavlov does not lead to this. **C** is incorrect: the bell on its own does not produce a naturally occurring unconditioned response, so it cannot be the unconditioned stimulus. **D** is incorrect because the sound of the bell is the neutral stimulus: it produces no predictable naturally occurring response.

Question 6

A the conditioned response.

The sound of the bell had been neutral: it did not produce any naturally occurring response. But after being paired repeatedly with the meat powder, the bell alone produces the salivation. Therefore, the salivation to the bell has been conditioned and it is a conditioned response. **B** is incorrect because an unconditioned response is a naturally occurring response and therefore unlearned. In this case, salivating to the meat powder would be the unconditioned response. **C** is incorrect because the conditioned stimulus was previously the neutral stimulus, so in this case the bell is the conditioned stimulus. **D** is incorrect because while it is correct that salivating to the bell is the conditioned response, the unconditioned response is a naturally occurring response and therefore unlearned. In this case, salivating to the meat powder would be the unconditioned response.

Question 7

A stimulus discrimination.

Stimulus generalisation happens when the conditioned response occurs in response to other stimuli. When it does not, it is called stimulus discrimination. **B** is incorrect because stimulus generalisation happens when the conditioned response occurs in response to other stimuli, similar to the original conditioned response. For example, if the dog was to salivate to the sound of a similar bell. **C** is incorrect because spontaneous recovery is the re-emergence of a conditioned response when the conditioned stimulus is once again presented after extinction and a rest period. **D** is incorrect because extinction is when the conditioned stimulus is presented without the unconditioned stimulus until the conditioned response no longer occurs. For example, Pavlov would ring the bell and not present any meat powder until the dog no longer salivates to the sound of the bell.

Question 8

A behaviour with desirable consequences is likely to be repeated.

Operant conditioning relies on behaviour being shaped by consequences. If consequences are positive, or desirable, the behaviour will be repeated. If they are not desirable, the behaviour is less likely to be repeated. **B** is incorrect because behaviour that is punished – such as the removal of a desired stimulus or the application of an undesirable stimulus – is less likely to be repeated. **C** is incorrect because behaviour that leads to a desirable consequence is more likely to be repeated. This is known as reinforcement. **D** is incorrect because behaviour that is punished is less likely to be repeated.

Question 9

D negative punishment.

Punishment decreases the likelihood of a behaviour occurring. Becki's phone being taken away is negative punishment, involving removal of a desired stimulus. **A** is incorrect because positive reinforcement involves the delivery of a pleasant or desired stimulus. In this case, a desired stimulus has been taken from Becki. Positive reinforcement makes it more likely that a behaviour will be repeated. **B** is incorrect because negative reinforcement involves the avoidance of an undesirable stimulus, making it more likely for the behaviour to reoccur. The aim of removing the phone is to make it less likely for Becki to send offensive texts in the future. **C** is incorrect because positive punishment is the delivery of an undesirable stimulus, such as a verbal reprimand. It will make it less likely for the behaviour to occur again.

Question 10

C every time a response occurs.

When conditioning, the link between the behaviour and consequence must be established. Therefore, it should be given consistently to establish the pattern. Once it is conditioned, the timing can change to strengthen the behaviour. **A** is incorrect because if reinforcement is given intermittently, it makes it difficult for the learner to establish the link between the behaviour and consequence. **B** is incorrect because if reinforcement is only given every second time a response occurs, it makes it more difficult for the learning to establish the link between the behaviour and consequence. **D** is incorrect because for the link to be established between the behaviour and consequence, the consequence should only occur after the behaviour/response has occurred. Otherwise, the learner will not establish the link between the behaviour and consequence.

Question 11

B Punishment must be harsh.

For punishment to be effective, it must be issued immediately after the behaviour occurs and it has to be something that is actually unwanted. Otherwise, the effect may be the opposite of that intended. The harshness or severity of a punishment does not contribute to its effectiveness. In fact, punishment that is too harsh can have a negative impact: the punishment can be linked to the person inflicting it, and not the behaviour itself. **A** is incorrect because the consequence needs to follow the behaviour for the link to be established between the behaviour and the consequence. **C** is incorrect because the punishment should immediately follow the behaviour for the punishment to be effective and association to be formed between the behaviour and consequence. **D** is incorrect because a person must find the consequence undesirable for it to be considered a punishment. This ensures that the behaviour is less likely to be repeated.

Question 12

B receiving a reduction in a jail sentence

A negative reinforcer is a good thing: it strengthens desirable behaviour by removing something bad. Therefore, reducing jail time is the only appropriate answer. **A** is incorrect: receiving a cheque for $200 is the application of a desirable stimulus, so it is positive reinforcement. **C** is incorrect: receiving a detention is a punishment, as it works to make the behaviour less likely to be repeated. By contrast, reinforcement makes a behaviour more likely to be repeated. **D** is incorrect: receiving a chocolate bar is the application of a desirable stimulus, so it is positive reinforcement.

Question 13

D All of the above.

All of the options take place in spontaneous recovery. The conditioned response must be extinguished; then there is a rest period, and at the presentation of the conditioned stimulus, the conditioned response reappears.

Question 14

B The class will behave well more often.

Operant conditioning works under the assumption that behaviour is shaped by its consequences. Therefore, the behaviour of the class should improve with appropriate reward and punishment in place. **A** is incorrect because Ms Lanati is using reinforcement every time her class is well behaved, therefore increasing the likelihood of her class behaving well in the future. Further, she is using punishment when the class misbehaves, making it less likely that the class will behave poorly in the future. **C** is incorrect because Ms Lanati is providing a consequence every time the students behave well (reinforcement) and when they behave badly (punishment). Therefore, the students should establish the link between the behaviour and consequence relatively quickly and permanently. **D** is incorrect because operant conditioning is based on the premise that behaviour becomes controlled by its consequences and can be shaped.

Question 15

C response cost, because having her lease terminated would result in Hannah losing her apartment

Hannah has removed the negative events (the loud parties), therefore this is response cost or negative punishment. Response cost must have a loss of something negative. **A** is incorrect because while response cost is the correct principle, response would involve the removal of a valued stimulus for Hannah (e.g. her apartment), not her neighbours being upset. **B** is incorrect because negative reinforcement is a principle which strengthens a behaviour. In this case, Hannah is less likely to repeat the behaviour of loud parties. **D** is incorrect because negative reinforcement is a principle which strengthens a behaviour by avoiding an undesirable stimulus. Were Hannah to receive an unpleasant consequence, this would be considered punishment.

Question 16

D being given compliments by her neighbours when she has a quiet party

The key word in this question is 'reinforce'. Compliments are a great example of reinforcement that increases the likelihood of a desirable behaviour reoccurring. **A** is incorrect because reinforcement makes a behaviour more likely to be repeated in the future. The police coming to Hannah's house should make it less likely that she hosts loud parties. **B** is incorrect because by ignoring the loud parties, there is no association being formed between Hannah's quiet and considerate behaviour and receiving a positive consequence. **C** is incorrect because her neighbours waiting outside develops no association between Hannah's quiet and considerate behaviour and receiving a positive consequence.

Question 17

D observational learning

Social learning theory is another term for observational learning, in which you learn through watching others. **A** is incorrect because operant conditioning is a different theory of learning, which suggests that behaviour becomes determined by its consequences. **B** is incorrect because classical conditioning is also a different theory of learning. It involves a previously neutral stimulus coming to elicit a reflexive response through repeated association with an unconditioned stimulus that produces a reflexive response. **C** is incorrect because childhood learning is not a learning theory.

Question 18

C reproduction

Although Jason is watching and retaining the information, and has great motivation as he wants to be a champion himself, he is only 10 and is unlikely to have the physical skills or strength to reproduce elite snowboarding. **A** is incorrect because Jason would have the ability to actively watch and pay attention to videos of snowboarders. **B** is incorrect because Jason would have the ability to create mental representations and retain the information he is watching. **D** is incorrect: Jason has the motivation to demonstrate this behaviour, as he wants to be a champion and snowboard like his idols.

Question 19

B a model acting aggressively.

According to observational learning, people replicate behaviours they observe, and are also guided by the consequences they see delivered. **A** is incorrect because Bandura found that after a child observed a model acting pleasantly, the child was more likely to imitate this behaviour, in accordance with the theory that people will replicate the behaviour that they see. **C** is incorrect because based on observational learning, a learner is likely to imitate the behaviour they see. Therefore, if a model does nothing, it is likely the learner will do nothing too. **D** is incorrect because by observing 'all of the above', the learner would imitate all three models – acting pleasantly, acting aggressively and doing nothing.

Question 20

A situated multimodal systems

The situated multimodal systems approach reflects the distinctive ways of knowing in Aboriginal and Torres Strait Islander cultures that situate the learner within connections to family, community and Country, and the multiple modes of cultural expression used to convey knowledge (including story, song, dance and art). **B** is incorrect because behavioural approaches to learning include only the processes of classical and operant conditioning and reflect a particular Western understanding of learning. **C** is incorrect because the social-cognitive approach focuses on learning through observation of a model and is another Western approach to learning. **D** is incorrect because observational learning is a form of social learning and involves five key processes (attention, retention, reproduction, motivation and reinforcement) and is a component of the social-cognitive approach to learning.

Short-answer sample responses

Question 1

Operant conditioning is a form of learning in which an association is made between a behaviour and its consequences to shape future behaviour. (1 mark) Its three stages are as follows.

- Antecedent: That which precedes the response; the event that stimulates the response to occur. (2 marks)

- Behaviour: The response by the person. (2 marks)

- Consequence: That which is issued following the behaviour to guide behaviour, either to reinforce (increase likelihood of behaviour occurring again) or punish (decrease the likelihood of the behaviour occurring again). (2 marks)

Question 2

In classical conditioning, the response is involuntary. (1 mark) In operant conditioning, it is voluntary. (1 mark)

In classical conditioning, the stimulus appears before the response. (1 mark) In operant conditioning, the response (behaviour) happens before the stimulus (consequence). (1 mark)

Question 3

Answers could set out a range of scenarios along the following lines.

- Attention: A child could have actively watched their parents cooking the evening meal together each night. (2 marks)

- Retention: They then hold a mental image of the steps involved in cooking. (2 marks)

- Reproduction: When they get the opportunity, they then have the physical skills and mental capability to be able to cook themselves. (2 marks)

Question 4

Situated learning occurs within authentic settings and situations where key knowledge is integrated within activity, context and culture. (1 mark)

For Aboriginal and Torres Strait Islander peoples, learning is situated on Country. (1 mark)

Country is central to the identities of Aboriginal and Torres Strait Islander peoples because it represents the spiritual, physical, social and cultural interconnectedness between all creation. (1 mark)

Test 4: The psychobiological process of memory

Multiple-choice answers

Question 1

C active

Memory is not a passive process: regardless of whether memory is formed implicitly or explicitly, a person is still active in their engagement. **A** is incorrect because memory is not a passive process – it involves actively processing, encoding and storing information. **B** is incorrect because encoding is an aspect of memory that involves putting information into a form in which it can be stored. **D** is incorrect because while memory can be complex, it is not known as a complex information-processing system.

Question 2

A converting information into a usable form for storage.

We attend to information as it moves from sensory to short-term memory. Once in short-term memory, it is encoded for storage. **B** is incorrect because encoding involves putting information into a usable form to be stored, not retrieved. **C** is incorrect because 'attending to information' is another way of saying paying attention to something, which is different to converting information for storage through encoding. **D** is incorrect because 'converting information' is a more correct way of describing encoding than 'attending to information'. Also, encoding is a process for storage, not retrieval.

Question 3

C short-term memory.

Short-term memory has a limited storage capacity of 5–9 items. **A** is incorrect because sensory memory has unlimited capacity. **B** is incorrect because the phonological loop is an aspect of short-term/working memory, not a separate store. **D** is incorrect because long-term memory has unlimited capacity.

Question 4

D episodic; declarative

Our long-term memory has two divisions; procedural ('knowing how') and declarative ('knowing that'). From our declarative memory branches episodic memory (autobiographical events) and semantic memory (facts and knowledge). **A** is incorrect because declarative memory is a branch of memory, and semantic memory is memory of facts and knowledge, rather than autobiographical or personal events. **B** is incorrect because semantic memory is memory of facts and knowledge, rather than autobiographical or personal events. **C** is incorrect because while episodic memory is the correct memory, semantic memory is a branch of declarative memory.

Question 5

B unlimited; 0.2–0.4 seconds

Although it is a storage system holding only 0.2–0.4 seconds, the capacity of iconic memory (as a sensory memory) is unlimited. Therefore, large amounts of information can be accessed, but are not held for long enough to be attended to. **A** is incorrect because capacity is the amount, or how much can be stored, not how long it is stored: the capacity is unlimited. The duration of iconic memory is not unlimited: it is very limited at 0.2–0.4 seconds. **C** is incorrect because the capacity is the amount, or how much can be stored, not how long it is stored: the capacity is unlimited. The duration is about how long information is stored, not the number of items that can be stored. **D** is incorrect because the capacity of iconic memory is unlimited. The duration is correct at 0.2–0.4 seconds.

Question 6

D declarative; procedural

Finding the bike was most likely an episodic memory, which is a branch of declarative memory ('knowing that'); knowing how to ride the bike is a procedural memory. **A** is incorrect because semantic memories are those of facts and knowledge: knowing where she stored the bike is Renee's personal memory and therefore it is more correct that it is a declarative memory. Knowing how to ride the bike is a procedural memory. **B** is incorrect because while knowing where she stored the bike is an episodic memory, knowing how to ride it is a procedural memory ('how to' memory). **C** is incorrect because procedural memories are our 'how-to' memories, not memories of information, such as where something is stored. Knowing how to ride a bike is a procedural memory, not a semantic memory (memory of facts and knowledge).

Question 7

A attended to.

From sensory to short-term memory, information must be attended to. This means we must pay attention to something for it to enter our conscious awareness. **B** is incorrect because encoding involves making information meaningful and putting it in a form to move it into long-term memory. **C** is incorrect because rehearsal involves elaborative rehearsal (adding meaning to the information). This allows information to be encoded into long-term memory. Maintenance rehearsal is another form of rehearsal that involves repeating information over and over: this allows it to remain in short-term memory. **D** is incorrect because retrieval refers to the process of bringing information from long-term memory to short-term memory.

Question 8

A how much information we can remember.

Capacity is how much information can be stored. **B** is incorrect because capacity is about how much can be stored, not what type of information can be stored. **C** is incorrect because how long information can be stored for refers to duration. **D** is incorrect because how quickly we forget information refers more to duration.

Question 9

A episodic memory

If Shantara is using information in her conscious control, it must be being manipulated in her short-term memory. The fact it is information from history is a strong indicator that this is not episodic memory for Shantara. **B** is incorrect because short-term memory is where we actively work with and manipulate information. Working hard to remember dates would definitely involve Shantara's short-term memory. **C** is incorrect because semantic memory is memory for facts and knowledge. Shantara is working with academic dates and facts; therefore, the memory store in question is semantic in nature. **D** is incorrect because declarative memory involves memories that can be 'declared', such as dates and other academic facts. Also, semantic memory is a form of declarative memory.

Question 10

D explicit memories.

The information is being consciously learned with the purpose of memorising it. Therefore, the memories are not implicit. And as the memories are semantic in nature rather than episodic, they must be explicit memories. **A** is incorrect because episodic memory deals with memory of personal experiences. The information being learned is semantic, involving facts and knowledge. **B** is incorrect because sensory memory is a brief store of an exact replica of incoming sensory information. We are not consciously aware of this information. **C** is incorrect because implicit memory is memory without needing conscious awareness. Learning theory requires conscious effort.

Question 11

D cerebellum

The cerebellum is focused on procedural memories or implicit memories. **A** is incorrect: while the hippocampus is vital for consolidation of memories, it does not store memories long-term. **B** is incorrect because the amygdala is responsible for attaching the emotional component to memories. **C** is incorrect because the neocortex is too broad to be the answer in this context.

Question 12

B   implicit memories, which are processed in the cerebellum.

Implicit memories are also known as automatic memories, and memories that are implicit in nature are most likely processed by the cerebellum. **A** and **C** are incorrect because explicit memories are not stored in the cerebellum. They involve episodic and semantic memories, not procedural memories such as how to ride a bike. **D** is incorrect because it is believed that implicit memories are consolidated and stored in the cerebellum.

Question 13

A   acronym

An acronym is formed by using initial letters to make a word. **B** is incorrect because an acrostic uses the first letter of a series of words or phrases to form a word. **C** is incorrect because method of loci involves linking each word to be remembered with a landmark, then retracing steps physically or mentally through those landmarks. **D** is incorrect because narrative chaining involves creating a story with the items to be remembered.

Question 14

C   method of loci

Jessica's technique involved linking the word to be remembered with a landmark, then retracing steps physically or mentally through those landmarks. This is known as method of loci ('locations'). **A** is incorrect because Chloe created an acronym by using the first letter of each stage to make a word. **B** is incorrect because an acrostic uses the first letters of a series of words or phrases to form a word. **D** is incorrect because narrative chaining involves creating a story with the items that need to be remembered.

Question 15

D   All of the above.

All options are advantages of method of loci. As you can continually add to the list of landmarks, method of loci can be used to recall large quantities of information in a particular order, and the landmarks act as a strong cue. Acronyms and acrostics can be used for ordered information, but can be limiting, as they need to make some sense verbally.

Question 16

C   keep the information in short-term memory.

The deep memory links made through song mean that information is in long-term memory, not short-term memory. **A** is incorrect because these songlines are extensively detailed and therefore create deep memory links. **B** is incorrect because these songlines are built and shared within communities. **D** is incorrect because these songlines are passed down through generations and therefore repeated.

Question 17

D   episodic memory

Not only does episodic memory have a place in retrieval of past memories, it also enables future memories to be projected by utilising those stored from past events and ideas. **A** is incorrect because amygdala store is not the name of a memory store. **B** is incorrect because short-term memory is limited in both capacity and duration. **C** is incorrect because sensory memory is a brief store of an exact replica of incoming sensory information; we are not consciously aware of this information.

Question 18

B aphantasia

Aphantasia is the inability to use visual imagery to project a mental image of oneself, a place or other people. **A** is incorrect because autism spectrum disorders are developmental disorders. **C** is incorrect because Alzheimer's is memory loss due to degeneration of neurons in the hippocampus. **D** is incorrect because aphasia is a language disorder that affects communication.

Question 19

B visual cortex

Your visual cortex processes information for your eyes, allowing you to make sense of what you see. It is also responsible for creating a visual image without the use of your eyes. **A** is incorrect because the basal ganglia plays a role in motor control. **C** is incorrect because the cerebellum plays a role in procedural/implicit memory consolidation and storage. **D** is incorrect because the amygdala is responsible for emotional memories.

Question 20

B It initially affects short-term memory more than long-term memory.

B is apt in describing the early stages of Alzheimer's. **A** is incorrect because the hippocampus is one of the first brain areas to be affected. **C** is incorrect because the main neurotransmitter involved in Alzheimer's is acetylcholine and a lack of this neurotransmitter. Some studies also suggest that a lack of dopamine is associated with Alzheimer's. **D** is incorrect because there are limited impacts on motor abilities.

Short-answer sample responses

Question 1 ©VCAA 2019 EXAM REPORT SB Q8 (ADAPTED)

Responses were marked holistically. Students are eligible for at least five marks if they:

- provided an adequate and accurate description of the Atkinson–Shiffrin model, including the function of the three stores and at least some aspects of their capacity and duration

- included a demonstration of their knowledge of the roles of short- and long-term memory (and related processes) within the model in **both** forming and retrieving explicit long-term memories

- related the above to at least **one** valid example from a person's memories of their first day at secondary school

- made some attempt to integrate at least one other relevant concept, theory or piece of evidence from Unit 3 within their response.

Responses may have included, but were not limited to:

- an outline of how the Atkinson–Shiffrin multi-store model of memory explains the flow of information in memory formation and retrieval with reference to function, capacity and duration of each store

- interactions between specific regions of the brain in the formation of a person's long-term memories of their first day at secondary school, including implicit and explicit memories. For example, if a person learned a new game of hopscotch – an implicit memory – the cerebellum would be involved in the formation and retrieval of that memory. However, learning new facts (explicit semantic memories) would be consolidated by the hippocampus and stored in the cerebral cortex. The amygdala would help consolidate emotional memories, such as those of making a friend, feeling nervous or being bullied on your first day

- other information from Units 3 and 4 connected with the formation and retrieval of a person's memories of their first day at secondary school.

Question 2

The neocortex is primarily involved in storing explicit memories, in particular semantic memories. (1 mark) Memories are transferred from the hippocampus to the neocortex for everyday applications. (1 mark)

Question 3

Written cultures or literary cultures utilise mnemonics that use words in print, such as acronyms or acrostics (1 mark), whereas oral cultures utilise mnemonics that use speech or song, such as songlines. (1 mark)

Students could provide a range of modern examples. One written example could be the use of the acronym DR ABC in CPR training. (1 mark) An example using song could be performing chest compressions during CPR to the beat of *Staying Alive*, as some first aid trainers recommend. (1 mark)

Question 4

Loss of episodic memories is common in sufferers of Alzheimer's disease. (1 mark)

Alzheimer's is brought about by amyloid plaques: proteins that accumulate in the brain, damaging and disrupting neural activity. (1 mark) Alzheimer's disease is also shown in the degeneration of neurons in the hippocampus. (1 mark) This highlights the importance of the hippocampus in the ability to consolidate short-term memories. (1 mark)

Test 5: The demand for sleep

Multiple-choice answers

Question 1

C an awareness of both internal and external events at any given moment

Consciousness can be internal and external. It involves how aware we are of these elements and our control of them. **A** is incorrect because consciousness is awareness both of internal events and external events. **B** is incorrect: awareness of how we feel is part of consciousness, but consciousness is also much more than that. **D** is incorrect because consciousness involves awareness of your internal and external world.

Question 2

B An altered state of consciousness can be naturally or purposely induced.

Although some altered states of consciousness are easy to interpret, others can be more difficult. Some are naturally occurring, such as sleep, while others are purposely induced, such as those induced by drugs. **A** is incorrect because it depends on a person's state: sometimes it can be difficult to determine whether someone is in an altered state of consciousness (e.g. distinguishing between a highly focused, yet altered state of consciousness, and normal waking consciousness). **C** is incorrect because everybody sleeps, and sleep is an altered state of consciousness. **D** is incorrect: consciousness is considered a psychological construct, as it cannot be physically observed and measured.

Question 3

B swimming

We are in normal waking consciousness when we swim. The other options are all altered states of consciousness. **A** is incorrect because being drunk involves changes in awareness, self-control and so on, all of which are characteristics of an altered state of consciousness. **C** is incorrect: sleep is considered an altered state of consciousness, as it involves marked differences from normal waking consciousness (e.g. a lower level of awareness). **D** is incorrect because meditation involves marked differences from normal waking consciousness (e.g. changes in awareness, time orientation etc.).

Question 4

B beta

Beta waves are associated with alertness; they are low-amplitude, high-frequency brainwaves. **A** is incorrect because alpha waves are associated with being awake but relaxed. **C** is incorrect because theta waves are associated with sleeping. **D** is incorrect because delta waves are seen when a person is in deep sleep.

Question 5

A an EEG

An EEG, or electroencephalogram, detects, amplifies and records electrical activity in the form of brain waves. **B** is incorrect because an EOG detects, amplifies and records electrical activity of the muscles that move the eyes. **C** is incorrect because an EMG detects, amplifies and records electrical activity of the muscles of the body. **D** is incorrect because a CT scan can be used to scan a brain, but does so by taking an X-ray-like image.

Question 6

D It would show no body movement.

An EMG detects, amplifies and records muscle movement. During REM sleep, a person is in a state of paralysis, so it would show no movement. **A** is incorrect because an EMG does not record brainwaves – an EEG does. **B** is incorrect because an EMG does not record electrical activity of the eye muscles – an EOG does. **C** is incorrect because a machine cannot show if a person is dreaming.

Question 7

D EEG; EOG; EMG

An electroencephalogram (EEG) detects, amplifies and records brain waves. An EOG detects, amplifies and records electrical activity of the muscles that move the eyes. An EMG detects, amplifies and records electrical activity of the muscles of the body. **A**, **B** and **C** are incorrect because the EEG should be positioned at number 1, the EOG at number 2 and the EMG at number 3.

Question 8

D while daydreaming

Daydreaming involves a person feeling relaxed: a state when alpha brainwaves predominate. **A** is incorrect because although alpha waves can be present, there are most commonly beta waves. **B** is incorrect: the brainwaves associated with sleepwalking are delta waves, as sleepwalking commonly occurs during deep sleep (stage 3 of NREM sleep). **C** is incorrect because the brainwaves that occur during NREM sleep are predominantly theta and delta.

Question 9

C hypnogogic state.

The hypnogogic state is experienced just before sleep onset and involves a person being very relaxed and having slow, rolling eye movements. **A** is incorrect because during REM sleep, beta-like waves are most likely to be evident, as well as bursts of rapid eye movement (not slow, rolling movements). **B** is incorrect because a hypnic jerk is a muscle spasm that happens most commonly in stage 1 NREM sleep. **D** is incorrect because there is no such thing as a theta state.

Question 10

C Usually the dreamer cannot remember the dream if woken.

C is correct as it is the only false answer. A person woken from a dream, especially a person in the REM state, is generally able to recall their dream upon waking. The other options are all true statements about dreams, and therefore incorrect. **A** is incorrect because dreams mainly occur in REM sleep. **B** is incorrect because when a person is dreaming and in REM sleep, their body does not move. **D** is incorrect because the brainwaves associated with dreaming and REM sleep are beta-like waves.

Question 11

B an electromyograph, which indicates changes in muscle tone

An electromyograph (EMG) records electrical activity in the muscles of the body. Therefore, it can help to show if there is a change in muscle tone due to movement throughout the night. **A** is incorrect because an EEG records electrical activity in the form of brainwaves. Therefore, it does not show if someone has moved during their sleep. **C** is incorrect. The key word here is 'quantitative'. While a video recording could show movement, the visual evidence is not a strong quantitative measure and is subject to individual interpretation. **D** is incorrect because an EOG records electrical activity in the muscles that move the eyes: this does not show if a person has moved in their sleep.

Question 12

A a biological cycle that lasts for approximately 24 hours

Our sleep–wake cycle is a great example of a circadian rhythm, as the cycle takes about 24 hours and then repeats. **B** and **C** are incorrect because a circadian rhythm is a biological cycle of approximately 24 hours, whereas biological cycles that occur more frequently are called ultradian rhythms. **D** is incorrect because a circadian rhythm is a biological cycle of approximately 24 hours; it is not a 1-month cycle.

Question 13

C urine production

Unlike the other three options, urine production does not follow a 24-hour circadian rhythm. It is triggered by many different factors such as hydration, diet and exercise. **A** is incorrect because the sleep–wake cycle is one of the best-known circadian rhythms. **B** is incorrect because body temperature is also a circadian rhythm, and is closely linked with the sleep–wake cycle. **D** is incorrect because hormone secretion is a circadian rhythm.

Question 14

B hypothalamus

The suprachiasmatic nucleus is found in the front of the hypothalamus and is involved in regulating the circadian rhythms of the body. **A** is incorrect because while the suprachiasmatic nucleus functions closely with the pineal gland, it is not found in the pineal gland. **C** is incorrect because the hippocampus is responsible for consolidation of memories (explicit). **D** is incorrect because the amygdala is responsible for emotional memories.

Question 15

D pineal gland; suprachiasmatic nucleus

Although the suprachiasmatic nucleus regulates our 24-hour circadian cycle and stimulates melatonin production as a result, it is the pineal gland that actually secrets the hormone. **A** is incorrect because melatonin is secreted by the pineal gland, not the hypothalamus. **B** is incorrect because the amygdala does not secrete melatonin; it is responsible for emotional memories. **C** is incorrect because the suprachiasmatic nucleus stimulates (activates) the pineal gland to release/secrete melatonin.

Question 16

B a sleep cycle

A sleep cycle is an ultradian rhythm as it occurs in a repeated pattern within a 24-hour period. Each sleep cycle is about 90 minutes long. **A** is incorrect, because the sleep–wake cycle runs on a 24-hour cycle, and is therefore a circadian rhythm. **C** is incorrect, because body temperature runs on a 24-hour cycle and is therefore a circadian rhythm. **D** is incorrect, because hormone secretion runs on a 24-hour cycle and is therefore a circadian rhythm.

Question 17

D Sarah 7 hours; Mia 16 hours

As an adult, Sarah needs about 7–8 hours of sleep each night. Newborn babies need about 16 hours or more, but this decreases as they get older. **A** is incorrect: 10 hours is the amount of sleep an adolescent requires, not Sarah, who is in her late 20s. As a newborn, Mia requires about 16 hours or more of sleep. **B** and **C** are incorrect because while Sarah needs about 8 hours of sleep, Mia needs about 16 hours of sleep.

Question 18

A 9; 20%

Teenagers need about 9–10 hours of sleep. About 20% of a typical night's sleep is spent in REM and 80% in NREM.

Question 19

B A higher percentage of NREM sleep is experienced towards the beginning of a night's sleep.

NREM sleep decreases as the night progresses, so we experience a higher percentage of NREM sleep at the beginning of the night. **A** is incorrect because the number of sleep cycles does not change much after a day of vigorous activity. At most, a person may sleep for a few extra minutes. **C** is incorrect because when we first fall asleep, we enter NREM sleep. **D** is incorrect because an average adult sleeps for about 8 hours, which is about five sleep cycles.

Question 20

B the adult, because adults have four to five sleep cycles per night

Although the duration of sleep is a little short for an adult, adults do typically have four to five cycles, as shown in the hypnogram. **A** is incorrect because the hypnogram does not show the time of sleep onset. Also, an adolescent would typically sleep for 9–10 hours. **C** is incorrect: while it is a true statement, spending 20% of sleep in REM is not unique to children. **D** is incorrect because this hypnogram does not show frequent waking throughout the sleep episode.

Short-answer sample responses

Question 1

A psychological construct is used to better describe human behaviour that would otherwise be difficult to effectively measure/observe the experience of. (1 mark) While there are markers to indicate someone is in an altered state of consciousness, these markers could also be due to other factors such as stress, making it difficult to accurately measure the experience. Hence, a psychological construct is needed. (1 mark)

Question 2

Theta waves have lower amplitude than delta waves. (1 mark) Theta waves have higher frequency than delta waves. (1 mark) A person may experience delta waves in stage 3 NREM sleep. (1 mark)

Question 3

a If patients had beta-like brainwaves, they would be in REM sleep (1 mark), and if they had theta or delta waves, they would be in NREM sleep. (1 mark)

Dr TiAni would use an EEG to measure this. (1 mark) If patients were moving their bodies, they would be in NREM (1 mark), and if they were completely still, they would be in REM. (1 mark) Dr TiAni would use an EMG to measure this. (1 mark)

If patients showed rapid eye movement, they would be in REM sleep (1 mark), and if they showed slow eye movement, they would be in NREM sleep. (1 mark) Dr TiAni would use an EOG to measure this. (1 mark)

b The proportion of time spent in REM sleep decreases from infancy to old age. (1 mark)

During infancy, there is an enormous amount of brain growth. As REM is responsible for replenishing the brain, more REM sleep is required earlier in life than in adulthood, when brain growth is not as marked. (1 mark)

Question 4

As Giorgio has been injured, his body may try to get more sleep, particularly NREM, which has disrupted his sleep–wake cycle. (1 mark) While he was in hospital, he may have been woken at different times or had no natural lighting, and so his sleep–wake cycle may have been disturbed. (1 mark)

This may have led to melatonin being released at times that were not in sync with his sleep–wake cycle. (1 mark) As a result, Giorgio will have difficulty regulating this cycle, perhaps feeling untired in the evening, or feeling tired during the day. (1 mark)

Test 6: Importance of sleep to mental wellbeing

Multiple-choice answers

Question 1

B reduced cognitive abilities

While a night's sleep deprivation hampers a person's ability to complete simple tasks, there is little impact on their ability to complete more difficult tasks, due to the necessity for concentration to be applied. **A** is incorrect because you would be more likely to experience low mood after sleep deprivation. **C** is incorrect because sleep deprivation tends to decrease our ability to concentrate and focus. **D** is incorrect because research suggests that performance on short complex tasks is not affected by sleep deprivation.

Question 2

B droopy eyelids; poor concentration

Physiological effects are those that affect the biology of the body (e.g. droopy eyelids). **A** is incorrect because poor concentration is a psychological effect, not a physiological effect. **C** is incorrect because irritability is a psychological effect, not a physiological effect. **D** is incorrect because while fatigue could be considered physiological, droopy eyelids cannot be considered psychological.

Question 3

D 0.10

Seventeen hours of sleep deprivation has similar effects to a BAC of 0.05, but after 24 hours of sleep deprivation, the effects closely resemble those of a BAC of 0.10. **A** is incorrect because studies show that the effects of being sleep deprived for 24 hours closely resemble those felt by a person with a BAC of 0.10 (double the legal driving limit). **B** is incorrect because the decimal is in the wrong place: it should read 0.10. **C** is incorrect because 17 hours of sleep deprivation has similar effects to a BAC of 0.05.

Question 4

A longer periods of REM

The body seeks to replenish what is lost, so a person deprived of REM sleep will likely have longer periods of REM sleep on subsequent nights, although it will still be close to the usual 80/20 split between NREM/REM. **B** is incorrect because the body aims to replenish what has been lost, and therefore will tend to need longer periods of REM sleep, not shorter periods. **C** is incorrect because a person typically sleeps slightly longer after sleep deprivation; they definitely would not be missing sleep altogether. **D** is incorrect because a person will still experience both NREM and REM sleep.

Question 5

A one night.

Lost sleep doesn't need to be made up in equal amounts. In fact, one night's good sleep is enough to reverse most of the side effects of sleep deprivation. **B**, **C** and **D** are incorrect because typically one good night's sleep is enough to replenish the body and mind after sleep deprivation.

Question 6

D microsleep.

A microsleep is a short burst of sleep in which a person appears to be awake. Microsleeps are often experienced by sleep-deprived people. **A** is incorrect because lacking concentration does not mean a person is having a period of sleep. **B** is incorrect because hallucination is experiencing something that is not there in reality. **C** is incorrect because a delusion is a strongly held belief that has no evidence to support it.

Question 7

D circadian phase disorders.

A circadian phase is a 24-hour cycle symbolic of our body clock or sleep–wake cycle. When that 24-hour cycle is disturbed, a person can experience a circadian phase disorder. Possible effects of this may include delayed sleep onset, insomnia or sleep disorders. **A** is incorrect because insomnia is a persistent difficulty with initiating and maintaining sleep, known as a dyssomnia. **B** is incorrect because delayed sleep onset involves a person falling asleep later than the average person, typically 2 or more hours later. This is a type of circadian phase disorder. **C** is incorrect because 'sleep disorders' is too broad in this context.

Question 8

C Think more coffee before a flight.

C is not helpful, as you are using stimulants, and not your natural body clock, to adjust your cycle. **A** is incorrect because natural light helps to reset the body clock. **B** is incorrect because it is good advice to try to keep the sleep component within a cycle at the right time, even if the length of the flight or sleep does not entirely match. **D** is incorrect. It is effective to start to create a shift, so the change is not as abrupt.

Question 9

A an ultradian rhythm within a circadian rhythm

The 90-minute cycles we have while asleep are ultradian. These cycles sit within the complete sleep–wake cycle, which is a circadian rhythm. **B** and **D** are incorrect because circadian rhythms run on a cycle of 24 hours and do not occur multiple times throughout the night. **C** is incorrect because the ultradian rhythm is a sleep cycle lasting for approximately 90 minutes.

Question 10

A   to reset the body clock

Bright light therapy uses the effects of natural light to stimulate feelings of awakeness and alertness and realign the sleep–wake cycle to the time of day. **B** is incorrect: while bright light therapy does help to make people feel more awake, **A** is a better description of its purpose. **C** is incorrect: while sore eyes can be a side effect of bright light therapy, this does not lead to sleep. The aim is to reset the body clock. **D** is incorrect because bright light therapy works by suppressing the release of melatonin and resetting its secretion.

Question 11

C   later; earlier

In delayed sleep phase disorder, a person does not feel tired at night-time, so they go to bed later. In advanced sleep phase disorder, people experience extreme sleepiness earlier than they would like to, so fall asleep earlier. **A** is incorrect: these terms are in the wrong order. **B** is incorrect because in advanced sleep phase disorder, people experience extreme sleepiness earlier than they would like to, so fall asleep earlier, not later. **D** is incorrect because in delayed sleep phase disorder, a person does not feel tired at night-time, so goes to bed later, not earlier.

Question 12

B   They wake too early from sleep.

As sufferers of advanced sleep phase disorder fall asleep early, they also tend to wake from sleep earlier than they would like to. **A** is incorrect because sufferers tend to fall asleep early, rather than feeling awake and alert later in the night. **C** is incorrect because struggling to breathe while sleeping is a common issue for people with sleep apnoea. **D** is incorrect because people with ASPD tend to fall asleep early and wake up early.

Question 13

B   Sleep hygiene is important to mental wellbeing.

While there is a known relationship between sleep and mental illness, there is not a causal relationship; that is, poor sleep does not cause mental illness. Instead, we know that good sleep habits, or sleep hygiene, leads to improved wellbeing. **A** is incorrect because there is no cause-and-effect relationship between lack of sleep and mental illness. **C** is incorrect because sleep problems can affect anyone. **D** is incorrect: Tom's teacher is not applying psychological knowledge by opening a conversation.

Question 14

C   Take an afternoon nap.

Taking an afternoon nap would not help Tom. Napping does not align with the 24-hour sleep–wake cycle. By contrast, a pre-bed routine (**A**), having a regular bedtime (**B**) and getting regular exercise (**D**) are effective strategies to improve sleep health.

Question 15

D   Electronic screens produce a lot of blue light, which can suppress the release of melatonin.

Blue light, emitted from electronic screens, suppresses the release of melatonin, the chemical that makes you feel sleepy. The sun also emits blue light, which is why we don't feel tired during the day. When the sun goes down, melatonin release is triggered. **A** is incorrect because blue light decreases at night, helping with the release of melatonin. **B** is incorrect because while blue light is emitted by electronic screens, the issue in terms of sleep is its suppression of melatonin. **C** is incorrect because Tom's teacher is concerned about the amount of blue light Tom is exposed to at night; she does not want to increase it.

Question 16

C 2–3 hours

Blue light should be avoided for 2–3 hours before bed as it suppresses melatonin production. **A** is incorrect because 30 minutes to 1 hour is not long enough to regulate the release of melatonin before bed. **B** is incorrect because 1–2 hours is not long enough to regulate the release of melatonin before bed. **D** is incorrect because studies show 2–3 hours is ideal; 3–4 hours is also unrealistic given demands including work, life and homework, which can all require screens.

Question 17

B Cues given by the external environment that regulate your circadian rhythm.

Zeitgebers are external cues that help regulate your circadian rhythm. They include things such as food intake at mealtimes, social interaction during the day and activity levels and patterns. The body uses these cues to regulate its natural 24-hour cycle. **A** is incorrect because zeitgebers are external cues, not internal cues. **C** is incorrect because these external environmental cues are helpful in the regulation of the sleep–wake cycle. **D** is incorrect because zeitgebers are external cues, not internal cues, that help – not harm – the sleep–wake cycle.

Question 18

A sunlight

Sunlight is more powerful than the patterns of eating and drinking (**B**), social interaction (**C**) and exercise (**D**). Not only is sunlight a very powerful and consistent cue, it also suppresses melatonin.

Question 19

B Our body temperature increases before bed, signalling to the body that it is bedtime.

Body temperature is a zeitgeber, as it is affected by things such as light cues and activity. It is at its highest around 6 p.m.: this is one of many cues for the body that it is time for bed. **A** is incorrect because body temperature is at its highest point before sleep. **C** is incorrect because body temperature is a circadian rhythm, meaning it runs as a 24-hour cycle; we would not have eight cycles within 24 hours. **D** is incorrect because body temperature is a zeitgeber.

Question 20

D All of the above.

Each of the statements in **A**, **B** and **C** are correct. We understand the importance of the need for sleep for good mental health; the psychological and physiological impacts of sleep deprivation; and the importance of light cues for the release of melatonin, which stimulates sleep.

Short-answer sample responses

Question 1

a A person who has had one night's sleep deprivation will have more extreme affective and cognitive effects than someone who has a BAC of 0.05, as the impacts of 24 hours without sleep are equivalent to the impacts of a BAC of 0.10. (1 mark) Affective effects may include increased anger. (1 mark) Cognitive effects may include decreased concentration. (1 mark)

b No. If the legal drinking limit was increased, people could legally drive while experiencing significant affective and cognitive effects. (1 mark) This would increase error, with drivers affected by lack of concentration, increased confidence, less attention to detail and worse judgement. (1 mark)

Question 2

Delayed sleep phase disorder can occur during adolescence, as there is a shift in the internal clock during puberty. This in turn results in the delayed release of melatonin. (1 mark)

Question 3 ©VCAA 2020 EXAM REPORT SB Q5

a   The 'body clock' refers to the circadian rhythm (1 mark) which regulates the sleep–wake cycle. (1 mark)

b   Melatonin is a hormone that induces sleep. (1 mark) Measuring melatonin levels will tell the researchers when the doctors are likely to feel tired. (1 mark) The researchers can then compare the melatonin levels to the data from the wrist band device to determine whether the light and movement data from the device accurately predict the onset of tiredness. (1 mark)

c   Affective functioning: A doctor might be snappy and irritable with a patient, which might affect the patient's willingness to speak about concerns. (2 marks)

Behavioural functioning: A nurse might be clumsy handling equipment and accidentally prick someone with a needle. (2 marks)

d   The device provides doctors with information about their body clock. (1 mark) The doctors could then use bright light therapy at times they need to be awake for a shift but their body clock is indicating they should be asleep, (1 mark) as exposure to bright light will delay the release of melatonin, which makes them feel sleepy. (1 mark)

e   Shift workers tend to work alternating rosters of day shifts and night shifts, which makes it difficult to adapt the sleep–wake cycle to match the work schedule, (1 mark) whereas for jet lag, individuals experience a single stable shift in the timing of light/dark to which they can adapt relatively quickly. (1 mark)

Test 7: Defining mental wellbeing

Multiple-choice answers

Question 1

B   increased resilience.

> Resilience is our ability to 'bounce back' after challenges or adversity. **A** is incorrect because 'better coping' is not a feature of a mental health problem; with a mental health problem, coping becomes more difficult. **C** is incorrect because resilience is our ability to 'bounce back' after challenges. The question references being 'better able to cope', suggesting increased not decreased resilience. **D** is incorrect because stigma refers to a mark or sign of shame: this question is not referring to issues of stigma.

Question 2

D   It helps to increase stigma.

> Increasing stigma would be a disadvantage, as stigma can affect people's willingness to seek help. Understanding mental health in terms of a continuum has been found to decrease stigma, because everyone sits somewhere along the continuum. All other options are an advantage of this approach, which helps to normalise mental health problems and disorders (**A**) by illustrating how people can have varying experiences (**B**) throughout life (**C**).

Question 3

C The severity of mental disorders can change over time.

People of all ages can experience mental disorders that are not necessarily related to trauma or life events. The experience can certainly change over time, including with the help of treatments. **A** is incorrect because mental disorders are not necessarily lifelong; they can change over time and can be effectively treated. **B** is incorrect because mental disorders are not necessarily due to external trauma; biological factors can contribute to the onset. **D** is incorrect because mental disorders often present early in life, such as in childhood or adolescence.

Question 4

B low resilience

Mentally healthy people have high resilience: they can 'bounce back' from challenges. **A** is incorrect because mentally healthy people tend to have low levels of anxiety, as they have the resources and ability to cope with stress. **C** is incorrect because mentally healthy people have high levels of social wellbeing and therefore strong connections. **D** is incorrect because mentally healthy people have high levels of emotional wellbeing and are able to recognise, understand and control their emotions.

Question 5

D functional

Many people who experience mental health problems do so without disruption to their day-to-day lives. Peta's approach aligns with the functional approach, which deems it necessary to seek help when everyday life is disrupted. **A** is incorrect because the medical approach to mental health looks at the physiological/biological factors that contribute to mental disorders and mental health. **B** is incorrect because the social approach to mental health looks at the social environment/context when explaining mental health and disorders. **C** is incorrect because the historical approach to mental health looks at how understandings of mental health and disorders have changed.

Question 6

A there is often an interplay of different factors contributing to a person's mental health.

At the core of the holistic approach is the ability to look at mental health from a biological, psychological and social perspective and consider the protective and risk factors in each of these areas. **B** is incorrect because the approach is holistic in terms of understanding a range of contributing factors from different domains, not a range of perspectives from different psychologists. **C** is incorrect: understanding that a person is a holistic entity is important, but 'holistic approach' refers to a comprehensive exploration and understanding of the factors contributing to mental health. **D** is incorrect: there are effective ways of coping with and treating mental health challenges.

Question 7

D All of the above.

Aboriginal and Torres Strait Islander peoples value looking at their health holistically, as they value their place in community, their connection to the land and their place within these systems (**B**). Contextualising each individual (**A**) brings richness to understanding mental health and wellbeing, and Aboriginal and Torres Strait Islander peoples are a population group with increased vulnerability (**C**) for historical and societal reasons.

Question 8

A Lily's teacher has applied a very narrow scope of mental wellbeing to Lily.

Lily's teacher has considered her mental health in biological terms; she has not considered the broad range of potential factors (i.e. she does not consider her mental health holistically). **B** is incorrect: the scenario suggests Lily is ready to seek help, as she is having an open conversation with her teacher. **C** is incorrect because Lily followed the advice, but it was not effective. The advice drew on a limited scope of wellbeing. **D** is incorrect because there is no indication the teacher is unqualified to discuss mental wellbeing.

Question 9

C a framework that represents a holistic approach to wellbeing for Aboriginal and Torres Strait Islander peoples, recognising the importance of culture and history in healthcare

Social and emotional wellbeing (SEWB) is a framework that strives to promote better mental health and SEWB outcomes for Aboriginal and Torres Strait Islander peoples, families and communities, by looking at SEWB holistically. **A** is incorrect because the framework applies broadly to many groups of Aboriginal and Torres Strait Islander peoples. **B** is incorrect because SEWB is a framework, not a plan. **D** is incorrect because the framework is for Aboriginal and Torres Strait Islander peoples living in all types of settings.

Question 10

C 7

There are seven overlapping domains: body, mind and emotions, family and kinship, community, culture, Country, and spirituality and Ancestors. **A**, **B** and **D** are incorrect.

Question 11

B connection to brain

The seven principles are body, mind and emotions, family and kinship, community, culture, Country, and spirituality and Ancestors. Connection to brain is not a principle. The other options are incorrect: connection to Country (**A**), connection to family and kinship (**C**), and connection to body (**D**) are instead important parts of the SEWB framework.

Question 12

A political

Political determinants are important to understand when exploring the framework; they allow us to consider the impact of government policies on the lives of Aboriginal and Torres Strait Islander peoples. **B** and **C** are incorrect because 'medical' and 'psychological' are not among the contextual determinants of SEWB. **D** is incorrect because 'sociocultural' is not one of the terms used to describe contextual determinants – social and cultural determinants are expressed separately.

Question 13

B because it integrates and connects individuals and communities with the past and future of their culture

A is incorrect because the phrase is not theoretical knowledge. **C** is incorrect because there are many ways that we learn about culture. **D** is correct in itself, but it does not reflect cultural continuity specifically.

Question 14

D exercise

Exercise is a protective factor. It is an important part of a healthy lifestyle and supports mental health. **A** is incorrect because substance abuse is a risk factor for the development of mental illness. **B** is incorrect because social isolation is a risk factor for the development of mental illness. **C** is incorrect because poor sleep is a risk factor for the development of mental illness. The other options are incorrect because substance abuse (**A**), social isolation (**B**) and poor sleep (**C**) are all risk factors for the development of mental illness.

Question 15

C stress, anxiety, phobia

This question is difficult, as it could depend on the severity of the condition and a person's level of functioning. However, stress is something we all experience, and it gives us the energy needed to confront a stressor. Anxiety can also be adaptive, as long as it is short-term. By contrast, a phobia is a completely irrational fear – a diagnosable mental disorder that severely affects a person's functioning. **A** is incorrect because a phobia significantly affects a person's ability to function, whereas stress and anxiety are considered normal and adaptive human responses. **B** is incorrect: stress should be considered the healthiest experience, as mild amounts can be adaptive and helpful for dealing with stressors. **D** is incorrect because while anxiety can affect a person's functioning to a great extent, if it is mild and short-term in nature, it can be considered adaptive. By contrast, a phobia is not considered adaptive and is a diagnosable mental disorder.

Question 16

C a stressor.

As it is the event that causes the stress, the date is the stressor. **A** is incorrect because stress systems are bodily systems, such as nervous systems, that produce stress responses. **B** is incorrect because a stress reaction would involve physiological and psychological responses to stress, such as increased heart rate and trouble concentrating. **D** is incorrect because stress is the state of physiological and psychological arousal produced by a stressor when a person perceives that they do not have the resources to cope with it.

Question 17

B a stress reaction.

Lauren's loss of appetite is due to her state of stress; therefore, it is a stress reaction. **A** is incorrect because the stressor is the stimulus that produces stress. In this situation, the stressor is the date. **C** is incorrect because stress systems are bodily systems, such as nervous systems, that produce stress responses. **D** is incorrect because stress is the state of physiological and psychological arousal produced by a stressor when a person perceives that they do not have the resources to cope with it.

Question 18

A low level of mental health concern

The stress Lauren is experiencing is perfectly normal, and may even be positive for her, in that it indicates she is showing her interest in the situation (eustress). **B** and **C** are incorrect: the stress Lauren is experiencing is normal, will be short-term in nature and will not affect her functioning. **D** is incorrect because Lauren is not suffering from a set of recognisable symptoms indicating a mental health disorder.

Question 19

B it is based on her belief about the outcome.

Psychological risk factors include affective, behavioural and cognitive impacts: Marguerite's beliefs are affecting her psychological state and therefore ability to cope. **A** is incorrect because her workplace would be considered a social risk factor. **C** is incorrect because expressing her concerns and talking them through may be considered a protective factor. **D** is incorrect because issues with neurohormones would be a biological risk factor.

Question 20

C emotional state; interactions with her boss

To discern the correct answer, we need to continue to refer to the scenario, in which we see Marguerite's emotional state (internal factor) affected by her interactions with her boss (external factor). **A** is incorrect because while physical health is an internal factor and family relationships are an external factor, physical health and family relationships are not issues identified in the scenario. **B** is incorrect because there is no suggestion Marguerite has a genetic predisposition to anxiety. **D** is incorrect because there is no suggestion of low self-esteem; also, conflict resolution skills would more likely be considered a protective factor.

Short-answer sample responses

Question 1

a   A mentally healthy person may be able to function in society and hold a regular job, (1 mark) whereas someone with a mental disorder may not be able to have regularity in their employment due to an inability to function consistently. (1 mark)

A mentally healthy person may have the resilience to be able to cope with a family misunderstanding, (1 mark) whereas someone with a mental disorder may not have the resilience to be able to reconnect with their family after a misunderstanding. (1 mark)

Other responses concerning social or emotional wellbeing would also be accepted.

b   No two people are alike in their ability to function in their environment, and so where one person may find their level of mental illness debilitating, another person with a similar experience may be able to cope and adapt. (1 mark)

Question 2

Resilience is our ability to cope with and manage change and uncertainty. (1 mark)

For example, during COVID lockdowns, many students were not allowed to attend school, and many important events were cancelled. (1 mark) Someone with high resilience may have been able to engage in online learning and overcome the disappointment of missing events within a day or so. (1 mark) Someone with low resilience may have struggled to overcome their sadness about cancellations and move forward. They may have found it difficult also to motivate themselves to learn online. (1 mark)

Question 3

A holistic approach to mental health is important, as people have potential in many areas, including physical, social, emotional and cultural. (1 mark) Looking at the health of all seven domains helps to ensure mental wellbeing, as all aspects of mental health are addressed and supported – not just one or two. If one domain is challenged by adversity, other domains can step up to play a greater role than usual. (1 mark)

A holistic approach can also be used in treatment to identify the greatest risks, as well as any strengths that can be utilised. (1 mark) For example, someone who is struggling physically and mentally may be able to draw on the resources of their community and on their connection to land to rebuild and reinvigorate their physical and mental health. (1 mark)

Question 4   ©VCAA   2018 EXAM REPORT SB Q7 (ADAPTED)

a   Students could respond with any of: mental health problem, mental health disorder or mental illness. (1 mark)

Students also gain one mark for effectively justifying their responses. For example:

- Students suggesting that Shari might be placed as experiencing a mental health problem could have justified this on the grounds that the severity of her symptoms did not seem to significantly affect her work and/or be of significant duration.

- Students suggesting that Shari might be experiencing either a mental health disorder or mental illness may have justified their response by noting that her symptoms seem to be worsening and that she is showing signs of distress, impacting on her social interactions (signs of paranoia in not trusting her colleagues), that she is distressed enough by her symptoms to have consulted a psychologist, and her problem has endured for a significant period of time.

b The following are examples of possible responses.

- Shari may be experiencing increased levels of stress since her company underwent restructure. (1 mark) Sustained stress can increase susceptibility to developing a mental health disorder by affecting the balance of neurotransmitters in the brain. (1 mark)

- Shari may have a genetic predisposition to developing a mental health disorder. (1 mark) Her underlying genetic predisposition could be exacerbated by the increased stressors at work since the restructure, increasing her susceptibility to develop a mental health disorder. (1 mark)

c Shari could increase her ability to adapt to stressful change at work through the use of cognitive behaviour therapy (CBT). (1 mark) Using CBT, the psychologist could identify Shari's persistent negative thoughts about her colleagues and ability to do her job and replace them with more realistic thoughts combined with behavioural change when the negative thoughts present. (1 mark) This could increase Shari's resilience as it could reduce the reoccurrence of stress response to changes at work. (1 mark)

Test 8: Application of a biopsychosocial approach to explain specific phobia

Multiple-choice answers

Question 1

B a phobia.

All key terms in the question are associated with the experience of a phobia. **A** is incorrect because stress is not considered irrational – in fact, it is quite an adaptive response. It gives us the energy to deal with stressors. **C** is incorrect because anxiety can sometimes be brief and not too intense. Also, anxiety can be general, and not in relation to a specific person, object or thing. **D** is incorrect because cumulative risk is the interaction of a range of risk factors that make it more likely for someone to experience a mental disorder.

Question 2

D anxiety disorder

As a phobia involves fear, and can be a projection of that fear in the future, it is an anxiety disorder. **A** is incorrect because mood disorders involve persistent changes in mood and interest in activities (e.g. major depressive disorder and bipolar disorder). **B** is incorrect because personality disorders involve long-term changes in thinking, behaviour and emotions. Behaviour and thinking tend to be rigid and dysfunctional (e.g. borderline personality disorder). **C** is incorrect because psychotic disorders involve a person losing touch with reality and often feature hallucinations and delusions.

Question 3

B It fails to inhibit a stress response.

GABA is an inhibitory neurotransmitter and therefore it is used to inhibit the stress response. In phobia sufferers, low levels of GABA can mean the response is not inhibited. **A** is incorrect because too much GABA is associated with having an increased inhibitory response in the CNS: it can be associated with daytime sleepiness, not anxiety and phobias. **C** is incorrect because GABA is an inhibitory neurotransmitter: it balances out the effects of excitatory neurotransmitters and reduces feelings of anxiety. **D** is incorrect because GABA does not block neural messages from reaching the brain. GABA is an inhibitory neurotransmitter and therefore has an impact on the brain and the neurons in it. GABA makes a postsynaptic neuron less likely to fire.

Question 4

A   an inhibitory neurotransmitter.

GABA is an inhibitory neurotransmitter and therefore it is used to inhibit the stress response. In people experiencing phobias, low levels of GABA can mean the response is not inhibited. **B** is incorrect because glutamate is the primary excitatory neurotransmitter; GABA is an inhibitory neurotransmitter. **C** is incorrect because GABA is a neurotransmitter and therefore a chemical, not a type of neuron. **D** is incorrect because GABA is a neurotransmitter and therefore a chemical, not a type of neuron.

Question 5

A   classical conditioning; operant conditioning

Phobias can often develop through the association of a fear response with a neutral stimulus, hence classical conditioning. When the sufferer avoids the stimulus they get a feeling of relief, which is a consequence like in operant conditioning. **B** is incorrect because operant conditioning is a form of learning whereby behaviour becomes determined by its consequences, whereas classical conditioning is a form of learning in which two stimuli become associated. Therefore, operant and classical conditioning are the wrong way around in this answer. **C** and **D** are incorrect because psychology is too broad an answer and biology is not relevant in this response. This question is looking at the forms of learning involved with the development of phobias.

Question 6

C   catastrophic thinking (psychological); psychoeducation (social)

The key to this question is to look at the classifications in brackets, as all can be considered contributing factors and evidence-based interventions. However, the classification must match. In this answer, catastrophic thinking is a psychological factor and psychoeducation is a social intervention. **A** is incorrect because while the stress response is a biological contributing factor, exercise is not a social intervention, it is a biological intervention. **B** is incorrect because while classical conditioning is a contributing factor, it is a psychological factor, not a biological factor. **D** is incorrect because while stigma is a contributing factor, it is a social factor, not a psychological factor.

Question 7

D   a negative stereotype

Stigmas are negative views about a particular thing. In this scenario, subscribing to the stereotype that men who suffer from mental health problems are 'weak' is perpetuating the stigma of mental health. **A** is incorrect because stigma is a negative view involving stereotyping and discrimination. **B** is incorrect because avoidance strategies are coping strategies that involve evading dealing directly with a stressor. **C** is incorrect because cognitive behavioural strategies involve dealing with the cognitive and behavioural aspects of mental disorders, which are not relevant to this question.

Question 8

A   education

One of the most effective ways of dealing with stigma is to educate and challenge people about their views. **B** is incorrect because punishment does not teach new ways of behaving; education is far more effective in reducing stigma. **C** is incorrect because while advertising could be an element of education, the best response is **A** (education). **D** is incorrect because helplines are more likely to assist people suffering from mental disorders. To reduce stigma, education of the wider community is needed.

Question 9

C They enhance the effect of GABA to reduce anxiety.

In a phobic response, GABA can be low, and therefore it does not inhibit the stress response. Benzodiazepines enhance the effect of GABA to inhibit the response and reduce the anxiety. **A** is incorrect because while benzodiazepines do enhance the effect of GABA, they work to reduce anxiety, not increase it. **B** and **D** are incorrect because benzodiazepines work by mimicking or stimulating the effect of GABA, and therefore enhance the effect of GABA and reduce anxiety.

Question 10

C They are an agonist, so they bind to the receptor to activate a response.

In a phobic response, benzodiazepines enhance the effect of GABA (agonist) to activate an inhibitory response and reduce anxiety. **A** and **B** are incorrect because an antagonist blocks neurotransmitter activity; in people with a phobia, we are trying to stimulate the activity of the neurotransmitter GABA. This is so GABA can bind to the receptor site to activate an inhibitory response. **D** is incorrect because while benzodiazepines are an agonist, they activate an inhibitory response and reduce the anxiety.

Question 11

D long-term potentiation.

As our neurons change, they form persistent patterns of response through the wiring of neurons through long-term potentiation. **A** is incorrect because GABA dysfunction refers to chemical imbalances in the brain. **B** is incorrect because memory bias is a psychological contributing factor, whereas neurological changes are biological factors. **C** is incorrect because long-term depression results in weakened synaptic connections due to lack of stimulation. By contrast, with a phobia, there are strong associations and connections in the brain: long-term potentiation.

Question 12

A biological.

The response involves physical changes to the brain's neurons: it is a biological response. **B** is incorrect because psychological factors include learning theories, memory biases and catastrophic thinking. The physical changes to neurons are a biological factor. **C** is incorrect because social factors include environmental triggers and stigma. The physical changes to neurons are a biological factor. **D** is incorrect because the physical changes to neurons are a biological factor, so social and psychological factors cannot be included in the answer.

Question 13

B specific environmental triggers and fear of stigma

Fear of judgement aligns with the concept of stigma, and the event itself is an environmental trigger. **A** and **C** are incorrect because although both are contributing factors, they do not align closely enough with the scenario. With classical conditioning, there generally needs to be repeated association, but this is a one-off event. **D** is incorrect because environmental triggers and fear of stigma are clearly present in the scenario.

Question 14

C social factors.

Environmental triggers and fear of stigma, as seen in the scenario, are both social factors. **A** is incorrect because biological contributing factors would include things such as GABA dysfunction and long-term potentiation. **B** is incorrect because psychological contributing factors would include things such as learning theories, memory biases and catastrophic thinking. **D** is incorrect because stigma and environmental triggers are social factors, so biological and psychological factors cannot be included in the answer.

Question 15

B catastrophic thinking.

The scenario is a clear demonstration of catastrophic thinking: expecting the worst without considering other outcomes. **A** is incorrect because stigma involves negative views about a particular thing. This is not shown by Lex's family or friends. Rather, it is her own perception and worries – her thinking. **C** and **D** are incorrect because Lex's experiences are now beyond stress or anxiety: she is clearly having irrational fears, characteristic of a phobia.

Question 16

A breathing retraining.

Breathing retraining is an evidence-based intervention effective in controlling the experience of phobic anxiety. **B** is incorrect because hyperventilating is rapid breathing, whereas Lex's parents are trying to get her to slow down her breathing. **C** is incorrect because although breathing can reduce feelings of anxiety, this is not a structure technique as described. **D** is incorrect because cognitive behavioural therapy uses cognitive techniques, such as challenging unrealistic thoughts, and behavioural techniques, such as not avoiding our fears.

Question 17

C focuses on challenging negative thought patterns; does not focus on challenging negative thought patterns

Cognitive behavioural therapy focuses on challenging negative thought patterns, whereas systematic desensitisation aims to replace a fear response with a relaxation response – it doesn't overtly challenge negative thought patterns. **A** is incorrect because cognitive behavioural therapy does not involve creating a hierarchy – only systematic desensitisation does. **B** is incorrect: the two responses are the wrong way around. Cognitive behavioural therapy does not involve classical conditioning; systematic desensitisation does. **D** is incorrect because breathing retraining can be used in both cognitive behavioural therapy and systematic desensitisation.

Question 18

D It can take time and needs a willing participant.

Cognitive behavioural therapy is an intervention that requires a number of sessions to be effective. The participant must be willingly involved, as it requires a change in their behaviour and challenges their thought patterns. **A** and **B** are incorrect because cognitive behavioural therapy is unlikely to lead to any harm. It is an evidence-based intervention that can reduce the experience of phobias. **C** is incorrect: a core principle of cognitive behavioural therapy is to ensure that techniques can be applied more generally.

Question 19

C a positive UCS.

The feared stimulus is the CS. To change this behaviour, we need to pair the conditioned stimulus with a UCS that produces a naturally positive response. **A** and **D** are incorrect because classical conditioning involves the association between a CS and a UCS. **B** is incorrect because systematic desensitisation aims to replace the fear response with a positive response, so a positive UCS is required.

Question 20

B increases in intensity.

Systematic desensitisation starts with a small amount of exposure to the feared stimulus, which is gradually increased during conditioning. **A** is incorrect because systematic desensitisation involves working through a hierarchy of fear, from least anxiety-provoking to most anxiety-provoking. Therefore, exposure increases in intensity. **C** is incorrect because systematic desensitisation is very structured and involves working through a hierarchy of fear. **D** is incorrect because the aim of systematic desensitisation is to remove a person's fear response by helping them to relax as they work through a hierarchy of fear, so that ultimately they no longer fear the phobic stimulus.

Short-answer sample responses

Question 1

The biopsychosocial approach explores the influence of biological, psychological and social factors on a person's mental health and, in the case of mental illness, their incidence of illness. (1 mark) It can be used to prevent mental illness by ensuring that there are protective factors in all three areas. (1 mark) It can be used in treatment to identify areas of concern or the best starting place to address, support and strengthen mental health. (1 mark)

Question 2

a   Dysfunction may be defined as an inability to interact socially, hold down a job or care for yourself. (1 mark) Some phobias can become generalised and affect sufferers' ability to function in society. By contrast, some experiences of anxiety can be well-managed and not compromise people's ability to function in society. (1 mark)

b   Psychoeducation (1 mark) helps family or supporters to challenge a sufferer's unrealistic or anxious thoughts. It also gives them techniques to get the sufferer to consider alternative outcomes and realities. (1 mark)

   Not encouraging avoidance behaviours is also an acceptable answer.

Question 3   ©VCAA   2020 EXAM REPORT SB Q6

a   Maxine's parents positively reinforced her running away from the red box (1 mark) by playing with her and comforting her. (1 mark) This reward increases the likelihood that Maxine will persist in avoiding similar red boxes into the future. (1 mark)

b   A benzodiazepine agent acts as a GABA agonist in the brain, (1 mark) promoting GABA's inhibitory effect, and so helps to calm physiological arousal. People with a phobia may have dysfunctional levels of GABA. (1 mark) Prescribing a benzodiazepine agent for Maxine would help reduce the extreme anxiety she experiences due to her phobia. (1 mark)

c   People with a specific phobia have a bias to recall the negative content of experiences with their phobic stimulus more than positive or neutral memories associated with it. (1 mark) Consequently, Maxine's negative memories of red boxes similar to the one that caused her extreme fear response are strengthened and/or are more likely to be recalled again in future. (1 mark)

Question 4

a   There is evidence of conditioning due to association between two stimuli. (1 mark)

b   The unconditioned stimulus is the cane toads jumping towards her (1 mark) and the conditioned stimulus is the cane toads. (1 mark)

c   This has occurred through stimulus generalisation. (1 mark) The conditioned stimulus of cane toads was similar to frogs, then the word reminded her of them and so on. In classical conditioning, there is a powerful association with the UCS and the CS, which can become generalised. (1 mark)

Test 9: Maintenance of mental wellbeing

Multiple-choice answers

Question 1

D at any time

Protective factors help to decrease the probability of mental illness occurring, and can be strengthened at any time. **A** is incorrect because protective factors can be developed and strengthened at any time to decrease the chances of developing a mental illness. Protective factors and interventions are particularly important when a person is suffering a mental illness, but should be developed regardless. **B** is incorrect because while it is important to develop and utilise protective factors when recovering from a mental illness, these protective factors can be developed and strengthened at any time. **C** is incorrect because while it is important to develop and strengthen these factors when mentally well to decrease the probability of developing a mental illness, they can be developed at any point – even when suffering a mental illness.

Question 2

B Clay has made a poor decision because this is an excellent time to learn more about the origins of his anxiety and help put strategies in place to prevent it from reoccurring.

Cognitive behavioural therapy takes time. It aims to work at the core of what is causing the anxiety. By stopping before he really understands or works on the source of his anxiety, Clay may regress when he is confronted with the stimulus in future. **A** is incorrect because cognitive behavioural therapy works very well when a person is anxious. It is an evidence-based intervention, but it also helps people understand their anxiety and how to implement cognitive and behavioural strategies to alleviate it and prevent its reoccurrence. **C** is incorrect because by learning to understand his anxiety and how to implement strategies to reduce its reoccurrence, Clay is developing an effective protective factor. **D** is incorrect because cognitive behavioural therapy helps Clay understand his anxiety and implement cognitive and behavioural strategies to lower his anxiety and help prevent its reoccurrence.

Question 3

C a genetic predisposition

A genetic predisposition involves having a genetic or biological vulnerability to developing certain disorders. **A** is incorrect because the experience of parents divorcing could be either a social or psychological influence. **B** is incorrect because feeling socially isolated is a social contributing factor. **D** is incorrect because having below-average intelligence is not strongly correlated with mental illness – and even if it were, it would be a psychological factor.

Question 4

B sleep is a biological protective factor.

Sleep is a basic biological need. Great sleep helps many functions, both physiological and psychological, and so is a strong protective factor. **A** is incorrect because adequate sleep is a biological protective factor, not a risk factor. Getting adequate sleep allows us to function at our best. **C** is incorrect because while sleep is a protective factor, it is a biological factor not a psychological factor. **D** is incorrect because adequate sleep is a protective factor not a risk factor. Further, sleep is a biological factor not a psychological factor.

Question 5

A Ensure you follow a good diet.

Health and nutrition are important protective factors for ensuring mental health and wellbeing. Emily should eat a well-balanced diet. **B** is incorrect because dieting with the aim of weight loss is not good advice: Emily needs to consume the right nutrients and energy to support her training, recovery and mental wellbeing. **C** is incorrect because Emily should also be prioritising her sleep to maintain her mental wellbeing. **D** is incorrect because this may help with Emily's swimming performance, but not her mental wellbeing.

Question 6

B psychological

Although stress can manifest in biological symptoms, it is a form of psychological tension resulting from perceptions, thoughts and experiences. **A** is incorrect because while stress can manifest in biological reactions and symptoms, it is caused by psychological factors such as thoughts and perceptions. **C** is incorrect because stress is not a social risk factor. Social risk factors include stigma, social isolation and so on. **D** is incorrect because 'mental' is not a risk factor. Risk factors are categorised as biological, psychological or social.

Question 7

D all of the above

The flexibility of mindfulness meditation means you can use it for a variety of purposes. **A** (increase patience and tolerance), **B** (reduce negative emotions) and **C** (focus on the present) are all applications, so **D** is the correct answer.

Question 8

B less likely to suffer from mental health problems.

No amount of preventative strategies can completely protect a person from mental illness. However, these strategies help to make mental illness less likely. **A** is incorrect: no amount of preventative strategies can completely protect a person from mental illness. **C** is incorrect because the more preventative strategies are used, the less likely someone is to suffer from mental health problems, so they are not at a greater risk. **D** is incorrect because preventative strategies play an important role in decreasing the risk of developing mental health problems.

Question 9

C cognitive; behavioural

These are techniques undertaken in cognitive behavioural therapy: a person has their thoughts or cognitions challenged and looks at including helpful behaviours in their daily practice. **A** is incorrect because challenging thinking is a cognitive strategy, not a social strategy. Social strategies would involve some form of social support. Getting out of the car each day is a behavioural strategy, not a biological strategy. **B** is incorrect because while changing your thinking could be considered psychological, it is an aspect of cognitive behavioural therapy and therefore more correctly identified as cognitive. Getting out of the car each day is a behavioural strategy, not a biological strategy. Biological strategies would involve things such as breathing retraining. **D** is incorrect because the behavioural strategy is getting out of the car, not adjusting thinking. Getting out of the car each day is not a social strategy, it is a behavioural strategy.

Question 10

D reframing.

Reframing involves challenging negative thoughts with more likely outcomes or more positive ways of viewing the scenario. **A** is incorrect because meditation involves focusing on one stimulus – often the breath – to quiet the mind and racing thoughts. **B** is incorrect because resilience refers to a person's ability to bounce back from adversity. **C** is incorrect because catastrophic thinking is a distorted form of thinking that involves assuming the worst. Chhaya is attempting to do the opposite of this.

Question 11

D stigma.

Stigma is a negative stereotype that can perpetuate inaccurate or unhelpful thoughts or beliefs. Unfortunately, mental illness carries stigma due to lack of understanding about sufferers, causes, symptoms and treatments. **A** is incorrect because stress is a state of physiological and psychological arousal caused by stressors whereby a person does not feel they have the resources to cope. **B** is incorrect because rumination involves repeatedly thinking undesirable thoughts without taking action to address them. **C** is incorrect because dissociation involves disconnecting from one's thoughts and feelings.

Question 12

B social

Friendships and connections are part of the social aspect of the biopsychosocial approach to mental health and wellbeing. **A** is incorrect because psychological protective factors involve a person's thoughts, feelings, perceptions and behaviours (e.g. having an optimistic outlook or using positive coping strategies). **C** is incorrect because biological protective factors involve factors such as limited genetic predisposition to mental disorders or adequate sleep and diet. **D** is incorrect because spiritual protective factors involve having a strong sense of meaning and purpose beyond oneself. This scenario emphasises Peter's social connections.

Question 13

A authenticity.

Authenticity is an important factor for establishing strong friendships, as is the energy that a friendship gives. **B** is incorrect because it is based on opinion, not scientific information. **C** is incorrect because this scenario does not suggest Peter is suffering from any type of stigmatisation. **D** is incorrect because resilience is generally attributed to individual people, not friendships.

Question 14

C the extent to which cultural contexts enable people to learn about and express their culture

'Cultural determinants' refers to the extent to which broader cultural contexts enable Aboriginal and Torres Strait Islander peoples to learn about and express their own unique cultures. **A** is incorrect: it is unrelated to psychological theory. **B** is too definite a statement: the development of mental illness is complex. **D** is unrelated to the SEWB framework.

Question 15

D cultural continuity

Cultural continuity and self-determination are two types of cultural determinants. **A**, **B** and **C** are terms unrelated to the theoretical knowledge used in SEWB.

Question 16

A may negatively impact mental wellbeing.

Transgenerational trauma and racism are among the reasons why Aboriginal and Torres Strait Islander peoples can be more susceptible to mental illness. **B** is incorrect: it is the opposite of the answer. **C** is incorrect because these are social or psychological influences, not biological influences. **D** is incorrect because these events are likely to have an effect.

Question 17

A our sense of connection to something bigger than ourselves, shaping our identity

Spirituality is about connection to something bigger than ourselves. It does involve connection to many things around us, but by definition relates holistically to life itself. **B** is incorrect because spirituality involves connecting to something beyond ourselves, and therefore beyond our minds. **C** is incorrect because while connection to Country does have spiritual meaning for Aboriginal and Torres Strait Islander peoples, this definition is too specific and does not cover the broad meaning of spirituality. **D** is incorrect because while connection to community does have spiritual meaning, this definition is too specific and does not cover the broad meaning of spirituality.

Question 18

B self-determination.

Self-determination is a powerful tool, allowing Aboriginal and Torres Strait Islander peoples to shape their own lives. **A** is incorrect because this is a different cultural determinant. **C** and **D** are not linked to cultural determinants.

Question 19

C It is passed from generation to generation.

One of the greatest strengths of healing practices in Aboriginal and Torres Strait Islander cultures is the passing of these traditions between generations over centuries. **A** is incorrect: journals publish peer-reviewed, academic research papers. **B** is incorrect because much Aboriginal and Torres Strait Islander knowledge is passed down through stories and cultural practices rather than through texts. **D** is incorrect because while practices can be specific to different communities, they are not thought of as 'stored', but shared.

Question 20

B Cultural continuity is maintained.

Cultural continuity is a key benefit of the way information about healing practices is shared. **A** is incorrect because not all people in a community become healers. **C** is incorrect because Aboriginal and Torres Strait Islander groups and cultures differ. There is no 'one-size-fits-all' healing practice. **D** is incorrect because there will always be people who suffer from mental health problems.

Short-answer sample responses

Question 1

Protective factors, such as a genetic disposition to being mentally well, do not prevent someone from developing a mental illness. (1 mark) Protective factors make developing a mental illness less likely by strengthening a person's ability to cope or move past adversity. (1 mark)

Question 2

The biopsychosocial approach explores the influence of biological, psychological and social factors on a person's mental health and, in the case of mental illness, on their incidence of illness. (1 mark) An advantage of the model is that it strengthens the value of psychological and social factors that exist alongside medical causes and interventions. (1 mark) A limitation of the model is that it suggests that mental illness cannot happen due to an isolated biological, psychological or social incident. Its integrated nature means that it is difficult to value independent influences. (1 mark)

Question 3

Biological factors: Diet and sleep are important factors for mental wellbeing. (1 mark) Terry has started to limit coffee, which should in turn help his sleep. (1 mark)

Psychological factors: Meditation is a great technique for reducing negative emotions. (1 mark) Terry has started to engage in meditation and wants to continue it in the future. (1 mark)

Social factors: Social supports and friendships that are authentic and energising are important for mental wellbeing. (1 mark) Terry calling friends and organising catch-ups is an important social factor. (1 mark)

Question 4

Cultural continuity: (1 mark) The ability to preserve historic traditions from a culture and carry them through generations into the future is important. (1 mark) It helps mental wellbeing by strengthening the bond to community and a person's connection to their culture and place within it. (1 mark)

Self-determination (1 mark): The fundamental right of Aboriginal and Torres Strait Islander peoples as the First Peoples of Australia to shape their own lives. (1 mark) It helps mental wellbeing by allowing people to live well according to their own values and beliefs. (1 mark)

Cultural continuity and self-determination are connected because cultural continuity depends on people being able to exercise their right to self-determination. (1 mark) While cultural continuity highlights the importance of the past, (1 mark) self-determination helps shape the future. (1 mark)

Test 10: Experimental design

Multiple-choice answers

Question 1

C the presence of meditation

The independent variable is what is different between the conditions – that which is being manipulated. In this scenario, that is whether the participants engage in meditation or not. **A** is incorrect because an independent variable is what is being manipulated, whereas the difference in scores between stress levels before and after is what is being measured. **B** is incorrect because the results after the experiment is what is being measured, not what is being manipulated. **D** is incorrect because time of day is not what is being manipulated. However, this could be considered an extraneous variable.

Question 2

A the difference in the stress rating reported before and after the experiment

The dependent variable is measured to ascertain the effect of the independent variable. In this scenario, the difference in stress rating before and after is what is measured. **B** is incorrect because 'the results after the experiment' is too broad; it does not explain exactly what is being measured. **C** is incorrect because the presence of meditation is the independent variable – the thing that is being manipulated. **D** is incorrect because this is not what is being measured. However, this could be considered an extraneous variable.

Question 3

D a controlled variable.

This was an element in the research that was created to control variables in this study. The experimenter does so to eliminate extraneous variables, but the 'week of no work' itself is a controlled variable. **A** is incorrect because an extraneous variable is any variable other than the independent variable that may cause a change in the dependent variable. In this scenario, not working for the week is aiming to reduce the presence of extraneous variables, and therefore is a controlled variable. **B** is incorrect because an outlier is a result that is significantly higher or lower than most others. **C** is incorrect because confounding variables are variables that systemically affect the dependent variable like an independent variable does. That has not occurred in this scenario.

Question 4

A every member of the sample having an equal chance of being in the control or experimental group.

Once a sample group is recruited, members need to be distributed in a fair and equal manner to the control and experimental groups. **B** is incorrect because this answer is referring to random sampling, not allocation. **C** is incorrect because we cannot give every member of the population equal chance of being in the experimental or control group: a person must be selected as part of the sample first. **D** is incorrect because the sample is drawn from the population, so the sample is already part of the population.

Question 5

B double-blind

This eliminates the placebo effect and the experimenter effect and therefore is double-blind. **A** is incorrect because a single-blind procedure is when either the participants do not know what group they are in, or the experimenter does not know what group the participants are in. **C** is incorrect because the procedure is known as 'double-blind', not 'double-placebo'. **D** is incorrect because 'Hawthorne' is referring to the Hawthorne Effect, in which participants modify their behaviour due to being observed.

Question 6

B stratified sampling

Convenience produces the least representative sample, whereas random sampling is more representative. Stratified sampling allows the experimenter to ensure each stratum is represented in the proportion they appear in the population. **A** is incorrect because while random sampling can provide a representative sample, it is not the most representative and systematic way to gain a sample from the population. **C** is incorrect because convenience sampling is least likely to produce a representative sample. It involves using people easily accessible to the researcher. **D** is incorrect: random allocation is not a sampling technique, but involves allocating members of the sample into either the control group or experimental group.

Question 7

B all patients at the hospital; the 40 patients in the study

The sample is the people actually used in research; they are drawn from the population, which is the group of interest. In this case, the group of interest is the hospital patients. **A** is incorrect because these options are around the wrong way. The population is the group of interest (all hospital patients) and the sample is the people drawn from the population (the 40 patients in the study). **C** and **D** are incorrect because the population is the group of interest (the hospital patients). This scenario specifies that Samantha wants to conduct a study for Glenvale Hospital. Therefore, the population cannot be all Victorians.

Question 8

C to determine the effect of the independent variable by comparing the results of the control group with the results of the experimental group

Control groups are extremely important, as they provide a basis for comparison so the effect of the independent variable can be measured. **A** is incorrect because the purpose of a control group is not to obtain more results, but to allow comparison with the experimental group to assess the effects of the independent variable. **B** is incorrect because while a control group is useful to assess the effects of the independent variable on the dependent variable compared to the experimental group, it does not directly measure this. **D** is incorrect because the control group does not allow for this: instead, this could be achieved with a repeated-measures experimental design or a matched-participants experimental design.

Question 9

A a confounding variable.

It is most certainly an extraneous variable, but as the research is trying to measure depression, the presence of a past diagnosis of depression could confound the results, so it may be a confounding variable. **B** is incorrect because the independent variable is the variable that is being manipulated; in this case, it is the time spent in hospital. **C** is incorrect because the dependent variable is the variable that is being measured. In this experiment, the dependent variable was the onset of depression, not whether a patient had depression or not to start with. **D** is incorrect because the placebo effect involves participants changing their behaviour based on the belief they are receiving an active treatment.

Question 10

D the type of model children viewed

The independent variable is what was manipulated, and subsequently different between the two groups. This was the type of model they watched. **A** is incorrect because the number of children included is the sample. **B** is incorrect because the number of aggressive acts is what was being measured: it is the dependent variable. **C** is incorrect because the intelligence of the children may have been a potential extraneous variable, but it was not what was being controlled by Bandura as an independent variable.

Question 11

D by adding up all the scores and dividing the total by the number of scores

The mean is a measure of central tendency, calculated by adding up scores in the data set then dividing by the number of pieces of data. This is also known as the average score. **A** is incorrect because the midpoint of the scores is known as the median. **B** is incorrect because the most commonly occurring number in the scores is the mode. **C** is incorrect because subtracting the lowest from the highest score gives you the range.

Question 12

A a matched-participants design.

Rick did a pre-test to determine stress levels then paired and placed participants based on this matched characteristic. Therefore, it is a matched-participants design. **B** is incorrect because a repeated-measures design involves using the same group of participants in all conditions of an experiment: they are both the control group and the experimental group. In this scenario, there are two different groups of participants. **C** is incorrect because while this is an experimental design, this answer is far too broad. The answer should specify the type of experimental design. **D** is incorrect because independent-groups design does not involve pre-testing and matching participants, then allocating one member of the pair to the control group and one to the experimental group.

Question 13

D order effects.

Rick is describing a repeated-measures design. One common extraneous variable associated with this is an order effect, whereby the sequence in which the two conditions are conducted can influence the results due to practice or boredom. **A** is incorrect because counterbalancing is a way to control for the disadvantage of order effects to which Rick is referring. **B** is incorrect because single-blind effects are not a disadvantage; rather, a single-blind procedure is a way of controlling some extraneous variables such as experimenter effects. **C** is incorrect because placebo effects involve participants changing their behaviour due to the fact that they believe they are receiving an active treatment.

Question 14

B secondary data

Literature reviews allow researchers to use others' findings; this is known as secondary data. This can be advantageous, as it reduces the time needed to conduct new research. **A** is incorrect because primary data is data that the researcher collects firsthand through an experiment. **C** is incorrect because subjective data is data that is from a participant's perspective, such as thoughts, feelings, self-report data and so on. **D** is incorrect because while a literature review may include forms of qualitative data, it could also include quantitative data.

Question 15

C It gives participants a chance to explain their behaviour.

C is the only true statement. The advantage of using qualitative data is the potential for detailed data. **A** is incorrect: qualitative data is not easy to compare, as it involves data such as individual perceptions, which is very hard to compare. **B** is incorrect because qualitative data is not numerical: that is quantitative data. **D** is incorrect because qualitative data can take considerable time to collect. This is due to it involving a participant's thoughts, feelings, perceptions and so on.

Question 16

A median

The median allows for more consistent data reporting, free from the influence of outliers, as the median is the midpoint of the data set. **B** is incorrect because the mode could potentially be affected if there is a repeated outlier outside of the main data set. **C** is incorrect because the mean can be significantly affected by outliers, due to the nature of calculating the mean. It is important because variance is not a measure of central tendency.

Question 17

C reliability

This question talks about replicating findings, so the researcher is checking they are reliable. **A** is incorrect because in order to have low sources of error, we want research to have low levels of variance when replicated with different samples and in different environments. **B** is incorrect because validity and reliability are often confused. Validity involves the study measuring what it claims to measure and the ability for the results to be generalised. **D** is incorrect: having high uncertainty is not relevant in this scenario, which is about reliability – being able to replicate results.

Question 18

D exercising his withdrawal rights.

Even after research is complete, the participant has the right to withdraw their results. This fits under the principle of withdrawal rights. **A** is incorrect because participants are allowed to withdraw their results, even after the study has been completed. **B** is incorrect because withdrawal rights are Alan's protected and ethical rights. **C** is incorrect because informed consent occurs before a study. The participant needs to be told of the true nature of the study and the risks involved, and provide written consent to take part.

Question 19

B voluntary participation.

The participants have freely chosen to take part; this is voluntary participation. **A** is incorrect because informed consent means the participant is told of the true nature of the study and the risks involved, and provides written consent to take part. **C** is incorrect because confidentiality refers to participants' right to have personal details and their involvement in the study kept private. **D** is incorrect because withdrawal rights mean the participant can leave the study at any time, and have their results removed, should they wish.

Question 20

C to enable researchers to meet their participants and to decide whether to use their results or not

If a researcher did use this time to decide about using results, this would be a major flaw in research. **A** is incorrect because this is an important part of debriefing. Debriefing allows for support and counselling to be made available if required. **B** is incorrect because this is an important part of debriefing. Debriefing allows for the researcher to explain the results to the participants and answer any questions. **D** is incorrect because this is an important part of debriefing. Debriefing allows for the researcher to answer any questions the participants still have about the research.

Short-answer sample responses

Question 1

a Independent variable: the parenting method (reward or punishment). (1 mark)

Dependent variable : the number of negative behaviours displayed. (1 mark)

b It is hypothesised that parents in Texas who use reward as their parenting style will see fewer negative behaviours in their children than those who use punishment as their parenting style. (2 marks)

c A possible extraneous variable could be the initial behaviour of the children, who may have been poorly behaved or well behaved before the experiment. (1 mark) This could mean that the poor behaviour in the punishment group was a product of their natural style, not the parenting. (1 mark) This could be eradicated by observing the children's behaviour before the experiment, then using a matched-participants design to ensure equal presentation of behaviour in each group. (1 mark)

Another extraneous variable could be that parents had to report on the negative behaviours themselves. (1 mark) This may have led to under-reporting: parents may have been embarrassed to report negative behaviours. Discerning what these are is also somewhat subjective. (1 mark) This could be eradicated or reduced by having the researcher watch and record the behaviour in the final 3 days. (1 mark)

Question 2

The aim of this research is to study the change in neurons when a memory is formed. (1 mark) The most appropriate way to sample would be via stratified sampling. (1 mark) The researcher could select one student of each age group and then ensure she has an equal number of boys and girls, giving her a representative sample. Alternatively, she could choose two students from each school. (1 mark)

Stratified sampling eliminates bias experienced through convenience sampling and the potential for random sampling to fail to secure a representative sample. For example, all members of a random sample may turn out to be male. (1 mark)

A case study is a detailed study of a small group of people or an individual person. (1 mark)

An advantage is that a lot of detail can be obtained. (1 mark)

A disadvantage is that it is difficult to generalise results to the broader population. (1 mark)

First, informed consent would need to be adhered to. (1 mark) Parents would need to sign a form indicating that they understand the risks involved and consent to the research. (1 mark) Withdrawal rights are very important when working with children. (1 mark) The researcher would need to ensure there is someone looking out for the children's welfare during her work (e.g. to take action if a child does not want to continue.)

Beneficence would also be a factor to consider.

Practice exam 1

Multiple-choice answers

Question 1

B pupils dilate, mouth dries, digestion is inhibited, heart rate increases

While necessary functions prepare the body for action, those that are not needed, such as digestion and bladder, are inhibited. **A** is incorrect because digestion is not stimulated when arousal is high. Digestion is not needed for survival when under immediate threat; energy is diverted to other parts of the body. **C** is incorrect because pupils do not contract; they dilate to let more light in and enhance vision. **D** is incorrect because bronchioles dilate rather than contract; this allows for more oxygen to the lungs.

Question 2

B A conscious response by the nervous system is voluntary and attention is given to the stimulus.

To find the correct answer, it's important to remember that conscious responses are voluntary, while unconscious responses are involuntary. A conscious response occurs within our conscious awareness, so attention is given to the stimulus. **A** is incorrect because conscious responses are not involuntary. **C** is incorrect because unconscious responses are not voluntary. **D** is incorrect because not all unconscious responses are unintentional. Often they are very intentional.

Question 3

A gamma-aminobutyric acid (GABA)

GABA is an inhibitory neurotransmitter that calms the body's anxious response. **B** is incorrect because epinephrine or adrenaline is a neurohormone in the context of VCE Psychology. **C** is incorrect because norepinephrine can be a neurohormone and neurotransmitter. It has an excitatory effect.

Question 4

D dopamine

Dopamine is known for the role it plays in motivation and reward. **A** is incorrect because in the context of VCE Psychology, we look at the role of glutamate as an excitatory neurotransmitter involved in memory and learning. **B** is incorrect because GABA is an inhibitory neurotransmitter. It makes it less likely for a postsynaptic neuron to fire and generally weakens a neural pathway. **C** is incorrect because serotonin is involved in regulation of the sleep–wake cycle and mood.

Question 5

B adaptive plasticity.

Our brains adapt to new experiences and situations throughout our lives. **A** is incorrect because developmental plasticity tends to only be present until early adulthood. **C** is incorrect because survival plasticity is not a type of plasticity. **D** is incorrect because a critical period is a fixed time in development in which exposure and learning must occur.

Question 6

A pupils contract

During the fight–flight–freeze response, our pupils dilate to let more light in so we can see better. **B** is incorrect because perspiration is likely to increase to cool down the body. **C** is incorrect: saliva production will decrease, as it is not essential for immediate survival. **D** is incorrect because heart rate will increase to allow more blood to be pumped around the body.

ANSWERS – EXAM 1

Question 7

B in the resistance stage of the General Adaptation Syndrome.

Sarah is feeling 'OK', so her body's resistance to stressors is probably above normal, despite her increased workload. **A** is incorrect because shock is not a stage of the General Adaptation Syndrome model. Instead, shock is part of the Alarm Reaction Stage of the model. Also, shock is when a person's resistance to the stressor drops below normal levels when first confronted with the stressor. Sarah is currently coping, so is not in shock. **C** is incorrect because when a person is in the exhaustion stage, they are no longer able to cope, their resistance to the stressor drops below normal levels and they are likely to experience more severe illnesses. Sarah is still coping, so it is more likely she is in the resistance stage. **D** is incorrect because Sarah appears to be coping and feels 'OK', so she is not having a nervous breakdown.

Question 8

B is above normal.

Sarah is in the resistance stage, in which the body's resistance to stress is above normal. **A** is incorrect because in the resistance stage, resistance to stress is above normal, not below normal. **C** is incorrect because stress that is above normal but then rapidly drops below normal is characteristic of exhaustion. **D** is incorrect because Sarah is in the resistance stage, in which the body's resistance to stress is above normal – it does not fluctuate above and below normal.

Question 9

C It supports changes in the body that help it respond to stress.

Cortisol increases things like glucose in the body and harnesses the ability to repair tissue damage. **A** is incorrect because cortisol allows a person to more effectively deal with stress over a long period of time by increasing glucose for energy needed to deal with stress. **B** is incorrect because cortisol is released to help a person deal with stress; it does not create stress. **D** is incorrect because cortisol does not eliminate adrenaline. Rather, high levels of adrenaline cannot be maintained throughout chronic stress, whereas high levels of cortisol can.

Question 10

C primary appraisal

Thania is evaluating whether this is a stressful situation or not – and if it is, is it harm, challenge or threat? **A** and **B** are incorrect because shock and countershock are components of the General Adaptation Syndrome model, which is a biological model of stress – it is not part of the Transactional Model of Stress and Coping. **D** is incorrect because secondary appraisal involves Thania considering the coping resources she has to deal with losing her job. This is more conscious and occurs after primary appraisal.

Question 11

D is unable to be researched experimentally because primary and secondary appraisals often occur simultaneously.

Although **D** is not a familiar limitation of the model, we can see that **D** is the most correct response by process of elimination. With all psychological theories, one of the common limitations is experimentation with subjective data. **A** is incorrect because the model does explain this. If a person feels their coping resources are inadequate, they will experience stress. **B** is incorrect because the model does explain why people can interpret events differently according to a number of psychological factors. **C** is incorrect because the model sees stress as a transaction between an individual and their environment.

Question 12

A avoidance strategy

Omah is avoiding work despite having time to complete it: he is using the avoidance strategy. **B** is incorrect because meditation generally involves focusing awareness on an internal stimulus, often the breath. **C** is incorrect because approach strategies involve dealing with the stressor directly, whereas Omah is avoiding dealing with the stressor. **D** is incorrect because the scenario does not reference exercise: Omah is simply avoiding doing his work.

Question 13

D The strategy is appropriate in the necessary context.

Strategies need to be flexible and appropriate for the context. Strategies need practise and should be easy to use. **A** is incorrect because to be most effective, a strategy should be easy to use. Otherwise, a person is unlikely to use it. **B** is incorrect because to be effective, a strategy needs to work when the person experiences stress. **C** is incorrect because to be effective, the strategy should have been well-practised and found to work for the person using it.

Question 14

C classical conditioning.

There is an association between Ros' favourite CD (the neutral/conditioned stimulus) and an unconditioned stimulus (dinner). **A** is incorrect because observational learning involves watching the actions and consequences of a model and imitating the behaviour. Ros is learning to associate two stimuli. **B** is incorrect because shaping is a form of operant conditioning. This scenario involves the involuntary behaviour of salivating, whereas operant conditioning involves voluntary behaviour. **D** is incorrect because operant conditioning involves voluntary behaviour, rather than the involuntary behaviour of salivating.

Question 15

A conditioned stimulus.

The song has been conditioned; it does not create a naturally occurring response on its own. It is the conditioned stimulus that now produces a conditioned response. **B** is incorrect because an unconditioned stimulus means it is unlearned, and automatically produces a reflexive response. The CD does not automatically produce salivation; Ros' salivation is a result of conditioning. **C** is incorrect because a conditioned response is a behaviour/response to the presence of a conditioned stimulus. A CD or a song is not a response, but a stimulus. **D** is incorrect because an unconditioned response is a reflexive response in the presence of an unconditioned stimulus. A CD or a song is not a response.

Question 16

C She is experiencing stimulus discrimination.

The new stimulus (the live recording) is not similar enough to the conditioned stimulus, so stimulus discrimination has occurred. **A** is incorrect because being hungry is more likely an antecedent, which is part of operant conditioning, and therefore not relevant to this question. **B** is incorrect: if stimulus generalisation had occurred, Ros would have salivated to the live recording (with the new stimulus, similar to the original conditioned stimulus, producing a similar conditioned response). **D** is incorrect because spontaneous recovery occurs after extinction and a rest period: this has not occurred in this scenario.

Question 17

D passive; active

Classical conditioning has an involuntary response, so the learner is passive. In operant conditioning, the response is voluntary, so the learner is active. **A** and **B** are incorrect because 'voluntary' and 'involuntary' refer to types of responses, not the role of the learner. **C** is incorrect because these are around the wrong way: in classical conditioning the learner is passive and in operant conditioning the learner is active.

Question 18

A has the motivation to put rubbish in the bin.

This is the best answer, as it is clear Frank knows how to put rubbish in the bin, he just does not want to. That is, he lacks motivation. **B** is incorrect because Frank is five years old – he should be developmentally ready to use a bin. It appears he is simply choosing not to. **C** is incorrect because Frank has observed this behaviour by his parents many times. **D** is incorrect because Frank has been able to store a mental representation. It appears he is choosing not to use the bin.

Question 19

B Children who viewed a model being punished showed fewer aggressive acts than those who saw a model receive no consequences.

The power of observational learning is also dictated by operant conditioning principles. When children saw the model punished, their aggressive acts decreased. **A**, **C** and **D** are incorrect because Bandura found that children who observed the model being punished showed fewer aggressive acts than children who observed the model receive no consequences.

Question 20

D median

Mean, median and mode are the three measures of central tendency. **A** is incorrect because variance does not measure central tendency; it measures the range of scores from lowest to highest. **B** is incorrect because standard deviation measures how dispersed the data is around the mean. **C** is incorrect because reliability refers to the consistency of findings.

Question 21

C it contains outliers, very small or large values in the scores that are not typical.

A mean can certainly swing due to outliers. Every score must count in the total that the mean is taken from. **A** is incorrect because the frequency of each score relates to a frequency distribution. **B** is incorrect because the mean is not necessarily affected by the range. **D** is incorrect because this is characteristic of the median.

Question 22

B whether a cause-and-effect relationship exists between two variables.

An experiment allows researchers to determine a cause and-effect relationship. **A** is incorrect because this is typical of a case study, not an experiment. **C** is incorrect because this is typical of an observational study, not an experiment. **D** is incorrect because this is typical of a literature review, not an experiment.

Question 23

D 5–9 items; 18–20 seconds

The key to answering this question correctly is attention to the wording: 'capacity' is how much, and 'duration' is a measure of time. **A** is incorrect because the numbers are around the wrong way. **B** and **C** are incorrect because capacity is 'how much' (not the number of seconds) and duration is 'how long' (not the number of items).

Question 24

A cerebellum

The cerebellum is involved with implicit memories such as motor control memories. The other three brain regions are associated with the formation of explicit memories. **B** is incorrect because the amygdala is involved in the consolidation of emotional memories and classically conditioned emotional responses. It does not store implicit memories. **C** is incorrect because the neocortex is involved in explicit memories. **D** is incorrect because the hippocampus is responsible for the encoding of explicit memories.

Question 25

B an acrostic.

The example is a sentence, so it is an acrostic. **A** is incorrect because an acronym is a pronounceable word, not a sentence. **C** is incorrect because the method of loci involves linking words to be remembered with landmarks, then retracing steps physically or mentally through those landmarks. **D** is incorrect because a songline refers to a memory technique used in indigenous cultures.

Question 26

A songlines

The word 'oral' is all about talking and singing. Oral cultures maximise the use of singing for memory. **B** is incorrect because acronyms are used in many cultures, and not just oral cultures such as in Aboriginal cultures. **C** is incorrect because acrostics are used in many cultures, not just oral cultures. **D** is incorrect because method of loci involves the use of mental imagery more than oral techniques.

Question 27

D REM sleep decreases in duration throughout the night.

REM sleep increases as the night goes on. **A** is incorrect: REM sleep is referred to as 'paradoxical sleep' as the brain appears highly active, yet the body is in a state of paralysis. **B** is incorrect because during REM sleep, the body is in a state of paralysis from the neck down. **C** is incorrect because the vast majority of dreaming occurs in REM sleep.

Question 28

C Their neural development slows down.

REM sleep is associated with replenishing the brain. The first years of sleep see enormous amounts of neural growth. As that decreases, so too does the proportion of REM sleep needed. **A** is incorrect because the first years of an infant's life involve significant amounts of neural growth, which then decreases as an infant moves into childhood. **B** and **D** are incorrect because while the rate of physical development remains high, NREM stage 3 is more associated with physical development than REM sleep.

Question 29

A an inability to sit still while listening to the music

The key term in this question is 'behavioural'. This response clearly highlights a behavioural impact. **B** is incorrect because a lack of interest is more an affective impact. **C** is incorrect because this is a cognitive effect of sleep deprivation. **D** is incorrect because this is not a known impact; given that it is a feeling, it would be considered affective.

Question 30

A sleep deprivation can result in poor cognitive functioning.

Phoenix's inability to remember the hair colour of the lead singer is a clear cognitive effect. **B** is incorrect because while this statement is correct, being unable to remember something is a cognitive effect, not an affective one. **C** is incorrect because this statement is not related to the problem with Phoenix's memory. **D** is incorrect because hallucinations are a rare side effect of more extreme sleep deprivation, and the scenario does not refer to Phoenix having hallucinations.

Question 31

C 0.10 and she might not stay within her lane on the road.

The equivalent blood alcohol concentration (BAC) is 0.10.

C was a stronger response than **A** because it referred to Phoenix's ability to concentrate, which is a BAC impact. **A** referred to droopy eyelids, which is not necessarily related to BAC impacts. **B** and **D** are incorrect because BAC 0.05 is equivalent to 17 hours' sleep deprivation, not 24 hours.

Question 32

C higher amplitude and lower frequency brain waves.

Phoenix's awareness has decreased, which is characterised by brainwaves that are higher in amplitude but occur less frequently. **A** is incorrect because Phoenix would be less aware, and therefore displaying less beta waves. **B** is incorrect because she would have lower levels of alertness. **D** is incorrect because she would have higher amplitude waves.

Question 33

D any of the above.

The suprachiasmatic nucleus (**B**) is in the hypothalamus (**C**), which signals the pineal gland (**A**) to release melatonin, so damage to any one of these areas may disrupt the sleep–wake cycle. (Hence, the answer is **D** – any of the above.)

Question 34

B circadian cycle.

Sufferers experience disruption to their circadian cycle, as they fall asleep earlier than they should during this cycle. **A** is incorrect because an ultradian rhythm is one that runs for less than 24 hours. Advanced sleep phase disorder affects the sleep–wake cycle and therefore affects the circadian rhythm. **C** is incorrect because the question refers to an advanced sleep phase disorder: a delayed sleep cycle is the opposite of this.

Question 35

A objective.

The data is objective because it is observable, either by measuring brain waves or oxygen rates. It is therefore quantitative data. **B** is incorrect because subjective data is open to interpretation and difficult to compare. It often involves a person's thoughts, feelings and perceptions. **C** is incorrect because qualitative data involves a person's thoughts, feelings and perceptions, not their physiological measurements. **D** is incorrect because the artificial nature of the environment is a disadvantage of sleep laboratories.

Question 36

C sleep hygiene.

Hope has great sleep hygiene habits, including a routine to help her sleep. **A** is incorrect because 'sleep starters' is not a psychological term. **B** is incorrect because zeitgebers are external cues that work with your circadian rhythm, such as sunlight. **D** is incorrect because 'circadian clocks' is not a psychological term.

Question 37

C Go to bed at the same time each night.

To get into a routine, it is important to go to bed at the same time each night (so **C** is correct) – not just when you start to feel tired (so **B** is incorrect). **D** is incorrect because light acts as a zeitgeber and can lead to the suppression of melatonin, the 'sleepy' hormone. Although **A** would still be helpful, screens should really be avoided for at least 2 hours before bed.

Question 38

A external; external

The death of Melissa's father and her financial strain are both external factors. Internal factors may be Melissa's ability to cope, her grief and so on. **B**, **C** and **D** are incorrect: both factors are external, as they originate outside of Melissa.

Question 39

B resilience.

Resilience is our ability to respond to change or challenge, and to 'bounce back'. **A** is incorrect because resistance is a stage of the General Adaptation Syndrome model (biological model of stress). **C** is incorrect because eustress is a positive psychological response to stress. **D** is incorrect because 'de-stress' is colloquial, and not a term explored in psychology.

Question 40

D They both involve feelings of worry.

Both anxiety and a phobia involve feelings of anxiety and worry. **A** is incorrect because anxiety and phobias can be experienced at any stage of the life span. **B** is incorrect because a phobia is a mental disorder, and anxiety can be a mental disorder, such as generalised anxiety disorder. **C** is incorrect because there are effective, evidence-based interventions for both that are highly effective at treating the causes and symptoms.

Question 41

B can be helpful in mild amounts.

As with stress, mild forms of anxiety can prepare the body for necessary action or improved performance. Phobias, by contrast, are not helpful. **A** is incorrect because both anxiety and phobia involve feelings of distress. **C** is incorrect because both anxiety and phobias trigger the fight–flight–freeze response. **D** is incorrect because both anxiety and phobias are influenced by a range of factors including biological, psychological and social factors.

Question 42

C environmental triggers

Environmental triggers are the social factor that can often be traced back to the evolution of a phobia. **A** is incorrect because cognitive biases are a psychological factor and involve a tendency to think in a way that involves errors and faulty decision-making. **B** is incorrect because catastrophic thinking is a psychological factor and involves overestimating and predicting the worst possible outcome when coming into contact with a phobic stimulus. **D** is incorrect because memory biases are a psychological factor and involve distortions of recollections of previous encounters with a phobic stimulus.

Question 43

D his behaviour is uncharacteristic and has had a negative impact on his wellbeing.

When a person is finding coping more difficult, they often behave atypically (out of character). **A** is incorrect because not paying bills is not in itself indicative of a mental health problem. **B** is incorrect because while both internal and external factors are contributing to Theodore's behaviour, this in itself does not mean he is suffering from a mental health problem. **C** is incorrect because internal and external factors are influencing Theodore's behaviour (e.g. internal factors such as his mood, and external factors such as losing his job). In addition, these factors in themselves are not indicative of a mental health disorder.

Question 44

D organising nutritious meals; challenging Theodore's negative thinking about the future; getting Theodore's friends to visit him regularly

D is better than **A** because improved nutrition is a biological strategy for improving resilience that a doctor could encourage. Providing medication, although a biological strategy, is less likely to be used initially to improve resilience. **A** is incorrect because while medication is a biological strategy, it is unlikely to be prescribed to Theodore initially. If improving resilience is the goal, nutritious meals are more likely to be implemented first. **B** is incorrect because these strategies are under the wrong headings. Theodore's friends bringing him food is 'social', joining a seniors' club is 'social' and exercise is 'biological'. **C** is incorrect because genetics and nutrition are unlikely to have a link to resilience.

Question 45

B It has context-specific effectiveness and demonstrates coping flexibility.

Theodore's strategy was related to his situation (context), and in doing something differently, he showed coping flexibility. **A** is incorrect: while Theodore's strategy could be considered an approach strategy, as it is dealing directly with the stressor, it also demonstrates coping flexibility, as he tries something else after his first strategy is unsuccessful. **C** is incorrect because while Theodore's strategy has context-specific effectiveness, it will not necessarily help him avoid stressful situations. **D** is incorrect because this has nothing to do with his age, and Theodore is demonstrating coping flexibility not inflexibility.

Question 46

A Only the researcher knew who would receive the placebo.

A is the best response. The researcher is 'blind' to the allocation of the GABA agonist or the placebo to participants. **B** and **D** are incorrect: the research assistant, whose role is to interact with the participants and administer the treatments, needs to know who is receiving the GABA agonist, and who is receiving the placebo, so they can measure and record results accurately. **C** is incorrect because neither the researcher nor the control group must know who is receiving a placebo. People's expectations about the efficacy of the GABA agonist (and inefficacy of the placebo) could influence the results. In the case of the researcher, it could lead to 'experimenter effects', such as preferential treatment.

Question 47

C validity

> Validity is the marker by which to examine if the research is measuring what it is supposed to accurately. **A** is incorrect because outliers are pieces of data that are significantly higher or lower than the majority of results. **B** is incorrect because reliability is about being able to replicate results – ensuring they can happen time and time again, and in many different contexts. **D** is incorrect because 'ethics' refers to guidelines to ensure research is conducted in a way that protects participants from harm. This includes ensuring the researcher acts in a professional manner at all times.

Question 48

A confidentiality

> Data used must be de-identified to protect participants' privacy. **B** is incorrect because deception means participants do not know the true nature of the study. This can influence results. **C** is incorrect because debriefing occurs once the research has been completed. It involves making participants aware of the results and what they mean. Participants are also offered counselling if needed (especially if deception has been used). **D** is incorrect because informed consent is making participants aware of the nature of the study and of any risks involved, then gaining their consent for involvement.

Question 49

C standardised instructions and double-blind procedures

> Standardised instructions and use of a placebo are both methods to minimise the influence of extraneous variables. However, as double-blind procedures reduce extraneous variables more than single-blind procedures, **C** is more correct than **A**. **B** is incorrect because convenience sampling does not reduce extraneous variables – it adds to them (e.g. making it more likely that individual participant differences will affect the results). **D** is incorrect because counterbalancing helps to control order effects. It does not control for the experimenter effect, which demands a single-blind or double-blind procedure.

Question 50

D provide access to volumes of data that the researcher may not be able to gather.

> **D** is the best answer: secondary data from the internet can help a researcher gather large volumes of data that would otherwise be difficult to gather. **A** is incorrect: data on the internet will not necessarily have satisfied ethical guidelines. **B** is incorrect: data on the internet, although published electronically, may never have been published elsewhere, and is not necessarily reliable and valid. **C** is incorrect because the data may not involve a large sample, and therefore may not be representative of the population.

Short-answer sample responses

Question 1

a John's peripheral nervous system would have registered the sensation of the bug on his arm. (1 mark) Sensory information would then have been sent to the central nervous system for registering and processing (via sensory neurons). (1 mark) John's central nervous system would then send motor information (via motor neurons) to his peripheral nervous system, (1 mark) which would lift the hand and swat the bug. This action is controlled by the somatic nervous system in the peripheral nervous system. (1 mark)

b John may experience a spinal reflex. (1 mark) The sensory message would be intercepted in the spinal cord by an interneuron. (1 mark) The motor response would occur independent of the brain before information is transmitted to the brain. (1 mark)

Question 2

a Long-term potentiation (LTP) is the strengthening of synapses to improve neural efficiency. (1 mark) This can be characterised by increased dendritic spines or increased axon terminal branches. (1 mark)

b Developmental plasticity involves strengthening neural connections and pruning others, so utilises the power of LTP. (1 mark) It is important for our ability to effectively use functions that are more important or frequent, and for maximising our neural potential. (1 mark)

Question 3

Nassim is utilising the impact of dopamine. (1 mark) Dopamine is associated with motivation and reward. (1 mark) The sense of satisfaction Nassim gets from ticking things off her list is a reward, which makes her crave that feeling again and therefore drives her to get more done. (1 mark)

Question 4

The gut–brain axis involves communication between the central nervous system and the enteric nervous system. (1 mark) It is bi-directional: the communication goes both ways, as does the influence of one nervous system on the other. (1 mark)

Question 5 ©VCAA 2019 EXAM REPORT SB Q3

a The animal trainer creates a conditioned fear response to snakes by presenting the dog with a snake, the NS, followed by the electric shock, the USC (2 marks). Over repeated trials, the dog learns the association between the snake and the shock, producing a conditioned fear response to the snake when it is presented on its own (1 mark).

b Antecedent: the snake (1 mark)

Behaviour: avoiding the snake (not approaching it) (1 mark)

Type of consequence: positive reinforcement (with treat given for avoiding/not approaching the snake in the second method used in each training session) (1 mark)

c Any two of the following:

- Negative (unpleasant, painful, punishing) stimuli, like an electric shock, act as powerful unconditioned stimuli for training avoidance of the neutral stimulus because they produce a reflexive fear response that becomes strongly associated with the neutral stimulus, in this case the snake.

- Negative stimuli that cause a startle or pain response are strongly related to innate survival mechanisms and so make strong unconditioned stimuli to associate with a neutral stimulus that you want the learner to avoid.

- Learning can occur very quickly when negative stimuli that are related to survival, such as pain or shock, are used as the unconditioned stimulus.

- Negative stimuli engage powerful emotional learning via the amygdala and so create a strong association with the neutral stimulus.

Question 6

a The hippocampus and amygdala are found in the medial temporal lobe. (1 mark)

b Hippocampus: The hippocampus is involved in the formation or consolidation of memories, particularly episodic memories. (1 mark)

Amygdala: The amygdala is involved with attaching emotion to memory. (1 mark)

c While the hippocampus consolidates the memory, the amygdala adds the emotion to it, therefore helping to dictate what we may learn from an experience and the strength that is applied to this memory. (1 mark)

Question 7

a Country is a concept of place as a system of interrelated living entities, including the learner, their family, communities and interrelationships with land, sky, waterways, geographical features, climate, animals and plants. (1 mark) In Aboriginal cultures, knowledge can be stored in Country along songlines. (1 mark)

b They both store information connected to place. (1 mark)

Question 8

a When we consider future events, we draw on episodic memories we have from the past, utilising episodic memory. (1 mark) However, it is the framework of semantic memory that scaffolds the construction by using schemas and looking at plausibility of the construction. (1 mark)

b They may suffer from aphantasia. (1 mark) This prevents them from holding mental imagery, not just of future events but of past events too. (1 mark)

Question 9 ©VCAA 2018 EXAM REPORT SB Q5 (ADAPTED)

a Acceptable differences include, but were not limited to: (2 marks)

- REM duration becomes longer over the later sleep cycles, whereas NREM becomes shorter.

- NREM stage 3 is generally evident in the first couple of cycles, whereas REM occurs throughout the night.

- Overall time spent in NREM is greater than REM.

- Non-REM sleep has three stages and REM has one.

b The following are examples of possible responses: (2 marks)

- The hypnogram of a healthy adolescent would show more REM cycles and longer periods of time in NREM stage 3 compared to the hypnogram of an elderly person.

- The hypnogram of a healthy adolescent is likely to show a very similar proportion of REM to NREM sleep to that of an elderly person, but an elderly person is likely to have less time in NREM stage 3 than a healthy adolescent.

Question 10

a One similarity is that they are both cyclic, so they run on a timely cycle. (1 mark) A difference is that a circadian rhythm is 24 hours and an ultradian rhythm is less than 24 hours. (1 mark)

b Causes include late release of melatonin (1 mark) and lifestyle factors such as shift work, poor sleep hygiene and poor study habits. (1 mark)

c Bright light is used in the morning and then avoided in the evening to keep stimulation from light in-line with the sun. (1 mark) This stimulates the release of melatonin in the evening to help reset the circadian rhythm. (1 mark)

d Natural light helps the natural sleep–wake cycle as it stimulates the release of melatonin, which causes sleepiness in the evening. (1 mark) Blue light harms the natural sleep–wake cycle if used in the evening, as it suppresses the release of melatonin and hence delays sleep at the desired time. (1 mark)

Question 11

a Ten marks allocated across the following:

- It is hypothesised that Alzheimer's sufferers who are given a new drug will report a greater diminishment of symptoms than those who are given a placebo.

- The independent variable is the presence of the drug or placebo. The dependent variable is the score on the five-point scale used to report on the diminishment of symptoms.

- Random sampling was used to ensure every member of the population had an equal chance of being selected. This also ensured the sample was free from bias, as a sample of patients at different stages of Alzheimer's was obtained.

- A double-blind procedure was used: the experimenter and patients were unaware of who was receiving the placebo and who was receiving the experimental drug. This helped to ensure that there was not a difference in the patients' expectations due to knowing they were in the experimental group or not. It also ensured that there would not be 'experimenter effects', such as preferential treatment from the experimenters.

b The varying stages of the progression of Alzheimer's may have been an extraneous variable. (1 mark) Those further progressed may have been more resistant to the drug as their symptoms were more established. (1 mark) You could ensure you only use participants who have consistency in diagnosis (e.g. all having an initial diagnosis less than 6 months ago). (1 mark)

c As the drug is experimental, informed consent would be important to ensure the patients understood any risks or side effects. (1 mark) Informed consent from a legal guardian would also be necessary due to the level of awareness of Alzheimer's sufferers. (1 mark)

Question 12

A holistic approach to mental health is advantageous because of the many factors involved in protecting and contributing to mental wellbeing. (1 mark) This is particularly important to Aboriginal and Torres Strait Islander peoples because of the importance of their connection to culture as a determinant of wellbeing. (1 mark)

Any two of the following domains: body, mind and emotions, family and kinship, community, culture, Country, or spirituality and Ancestors. (2 marks)

Question 13

Biological factor: Explanation of GABA dysfunction or long-term potentiation. (1 mark)

Psychological factor: Explanation of the impact of classical and operant conditioning or cognitive biases. (1 mark)

Social factor: Explanation of an environmental trigger or stigma. (1 mark)

Practice exam 2

Multiple-choice answers

Question 1

D parasympathetic nervous system

> The body maintains homeostasis, controlled by the parasympathetic nervous system. When a threat is present, the sympathetic nervous system will activate. **A** is incorrect because the sympathetic nervous system activates when a threat is present, readying the body for action. **B** is incorrect, because 'peripheral nervous system' is too general a response for this question. **C** is incorrect because the somatic nervous system is responsible for control of voluntary skeletal muscle movements.

Question 2

D breathing

> Breathing is controlled by the autonomic nervous system. The other behaviours are all voluntary and therefore under the control of the somatic nervous system. **A**, **B** and **C** are incorrect because the somatic nervous system is responsible for the voluntary movement of skeletal muscles. Serving a tennis ball, opening a can of drink and playing soccer all involve voluntary movements and therefore result from the somatic nervous system.

Question 3

C Sensory information travels via sensory neurons to the spinal cord; it is intercepted by an interneuron, then a motor neuron takes motor information back to the site.

A spinal reflex involves all three types of neurons, which each execute their role as part of normal neural transmission, with the exception of the interception in the spinal cord for a spinal reflex. **A** is incorrect because motor information travels from the spinal cord to the muscles, not to the spinal cord. Sensory information travels to the spinal cord, not from the spinal cord. **B** is incorrect because this response does not refer to the interneurons in the spinal cord, which are responsible for intercepting the sensory information and relaying it to a motor neuron. **D** is incorrect because interneurons are only found in the central nervous system and therefore cannot take information from the body to the spinal cord.

Question 4

A pre-synaptic neuron; releases neurotransmitters from vesicles

This question involves a process of elimination. The presynaptic neuron is responsible for releasing neurotransmitter into the synaptic gap. **B** is incorrect because chemicals (neurotransmitter) travel across the synaptic gap, not electrical messages. **C** is incorrect because the receptor site does not store neurotransmitter. **D** is incorrect because reuptake happens at the presynaptic neuron.

Question 5

D adaptive plasticity.

As damage has been involved, the plasticity Aaliyah is experiencing is adaptive. She will continue to be aided by developmental plasticity along her journey. **A** is incorrect because a tumour would most likely contribute to issues rather than aid recovery. **B** is incorrect because while practice can lead to long-term potentiation, it alone does not lead to recovery. **C** is incorrect because developmental plasticity is present in childhood and adolescence, not into adulthood.

Question 6

B Glutamate is an excitatory neurotransmitter.

Glutamate stimulates a response in memory and learning, so is excitatory. **A** is incorrect because glutamate is a neurotransmitter, which is a chemical: it is not an electrical impulse. **C** is incorrect because glutamate is an excitatory neurotransmitter, while GABA is an inhibitory neurotransmitter. **D** is incorrect because glutamate is an excitatory neurotransmitter.

Question 7

D dopamine

Dopamine is associated with reward and pleasure. **A** is incorrect because glutamate is an excitatory neurotransmitter involved in learning and memory. **B** is incorrect because GABA is an inhibitory neurotransmitter involved in the calming of the central nervous system. **C** is incorrect because serotonin is a neurotransmitter involved in mood regulation.

Question 8

C stomach ulcers; anxiety

Physiological effects occur in the body (e.g. stomach ulcers). Psychological effects can include cognitive, affective and behavioural impacts (e.g. feelings of anxiety). **A** is incorrect: irritability is a psychological effect, as it is an affective impact (emotions and feelings). **B** is incorrect: both dizziness and headaches would be considered physiological impacts, as they are occurring in the body. **D** is incorrect because these are around the wrong way: forgetfulness would be a psychological impact, whereas heart palpitations are a physiological effect.

Question 9

D alarm reaction.

Alarm reaction stage is made up of both the shock and countershock stages discussed. The next stage will be resistance, in which the body stays above normal levels of resistance. **A** is incorrect because countershock is not a stage of the General Adaptation Syndrome, but a component of the alarm reaction stage. **B** is incorrect because exhaustion is the final stage of the General Adaptation Syndrome, when resistance to the stressor drops below normal levels. **C** is incorrect because resistance is the second stage of the General Adaptation Syndrome and involves the ability to resist the stressor staying above normal levels for a prolonged period.

Question 10

B increases glucose in the bloodstream and reduces inflammation

The most accurate statement is **B**, as the main role of cortisol is to increase the amount of glucose available for energy to deal with a stressor. **A** is incorrect because cortisol does not stop the immune system, but weakens it. **C** is incorrect because there is prolonged release of cortisol in the resistance stage of the General Adaptation Syndrome, which keeps a person's ability to deal with a stressor above normal levels. **D** is incorrect because adrenaline is released prior to cortisol.

Question 11

C coping flexibility; positive

Trying a new approach is an example of coping flexibility, which is a positive response to stress. **A** is incorrect because burnout is not present in this scenario. Also, changing a technique to better deal with a stressor is a positive approach, not a negative one. **B** is incorrect because approach strategies attempt to deal directly with the stressor. Changing to a more effective technique is an example of coping flexibility. **D** is incorrect because avoidance strategies involve not directly dealing with the stressor itself, but rather the negative emotions associated with the stressor. This is not relevant to this response.

Question 12

A A UCS produces a UCR.

The first stage is the naturally occurring response when an unconditioned stimulus (UCS) produces an unconditioned response (UCR) through no conditioning. **B** is incorrect because when a conditioned stimulus (CS) produces a conditioned response (CR), learning/conditioning has occurred, and therefore it is the 'after conditioning' phase. **C** is incorrect because a CS does not lead to a UCR. A UCR is an unlearned response that occurs naturally in response to the UCS. **D** is incorrect because initially a neutral stimulus (NS) produces no relevant response. However, after association with a UCS, the NS becomes the CS, which then leads to a CR.

Question 13

C the electric shock.

The electric shock is a stimulus that leads to a naturally occurring response. It is unconditioned/unlearned. **A** is incorrect because the light switch itself does not lead to a naturally and automatically occurring response. It is the electric shock that leads to the unconditioned response. **B** is incorrect because the light switches around the house initially do not produce a naturally and automatically occurring response. **D** is incorrect because sweating is a response, not a stimulus.

Question 14

D sweating at the sight of the light switch.

The conditioned response is Tara's reaction after conditioning has occurred. She now sweats at the sight of a light switch. **A** is incorrect because the light switch is a stimulus, not a response itself. **B** is incorrect because the light switches are a stimulus which produce a response, they are not a response in themselves. **C** is incorrect because the electric shock is the unconditioned stimulus, as explained in the previous question.

Question 15

B Tara's conditioned response will have been extinguished.

After repeated experiences without an electric shock present, her response will extinguish. **A** is incorrect because over the past 2 weeks, Tara will not have received an electric shock, and therefore should no longer sweat at the sight of a light switch. **C** is incorrect: an unconditioned response cannot be extinguished, as it is a naturally and automatically occurring response. **D** is incorrect because it is likely that the response will be extinguished, so Tara should have no ongoing fear of light switches.

Question 16

A negative reinforcement removes an unpleasant stimulus, whereas positive reinforcement gives a pleasant stimulus.

Both types of reinforcement strengthen behaviour. However, negative reinforcement does this by removing something negative. By contrast, positive reinforcement involves giving something positive. **B** is incorrect because negative reinforcement also increases the likelihood of a behaviour reoccurring. **C** is incorrect because negative reinforcement involves the removal or avoidance of an unpleasant stimulus, not the removal of a pleasant one. **D** is incorrect because both types of reinforcement strengthen behaviour/make it more likely to reoccur.

Question 17

C reflexive; voluntary

The response in classical conditioning is naturally occurring (reflexive), and in operant conditioning is voluntary, not reflexive. **A** is incorrect because these are around the wrong way. Classical conditioning involves a reflexive response, whereas operant conditioning involves a voluntary response. **B** is incorrect because classical conditioning does not involve a spontaneous response, and operant conditioning involves a voluntary response. **D** is incorrect because the response in operant conditioning is voluntary not reflexive.

Question 18

A Learning is linear.

Learning is not linear in these techniques, as it involves expansion and digressions in response to the learning. **B** is incorrect because learning does involve links to the land. **C** is incorrect because learning is linked to Country. **D** is incorrect because learning does involve visualisation.

Question 19

B people can remember limitless amounts of information.

Capacity refers to how much a person can remember, and **B** is the only response that refers to capacity. **A** is incorrect because the time a person can remember information refers to duration, not capacity. **C** is incorrect because there are a number of theories and explanations as to why forgetting occurs. **D** is incorrect. While the statement is true, it has nothing to do with the capacity of long-term memory.

Question 20

C Declarative memories are involved with 'knowing that'.

Declarative memory has two branches, semantic (dealing with facts) and episodic (dealing with events and experiences), and the learning is very explicit. **A** is incorrect because procedural memory is not a type of declarative memory, but a type of implicit memory. **B** is incorrect because declarative memory is also referred to as explicit memory. **D** is incorrect because declarative memories can be memories of facts (semantic memory), but also memories of personal events and experiences (episodic memory).

Question 21

D implicit memory.

Implicit means it has occurred without concerted, focused attention. **A** is incorrect because explicit memory is memory with conscious awareness and effort. Shien was surprised to find she knew all the words, suggesting the memory is implicit (without conscious awareness). **B** is incorrect because a photographic memory generally refers to a visual memory. **C** is incorrect because a semantic memory is a memory of facts and knowledge and requires conscious awareness.

Question 22

C the hippocampus

The hippocampus is responsible for memory consolidation and sends information to other brain regions for storage. **A** is incorrect because the reticular formation is involved with sleep and consciousness. **B** is incorrect because the cerebral cortex stores memories after they have been consolidated. **D** is incorrect because the corpus callosum is the structure that connects the two hemispheres of the brain.

Question 23

A cannot produce mental visual imagery.

People with aphantasia are unable to hold imagery as a mental representation. **B** is incorrect because personality changes do not have a label like this. **C** is incorrect because short-term memory loss generally has something to do with a brain injury that affects memory duration or capacity. **D** is incorrect because long-term memory loss is referred to as amnesia.

Question 24

D Alzheimer's disease is a neurodegenerative disease.

Alzheimer's is neurodegenerative, meaning it is a progressive illness with effects that will continue to worsen. **A** is incorrect, because although Alzheimer's affects mental processes and functioning, **D** is the best description. **B** is incorrect, because although Alzheimer's is more prevalent in older people, it is possible to develop early-onset Alzheimer's disease. **C** is incorrect because Alzheimer's is not a contagious disease that can be caught from others.

Question 25

B 6.30 a.m.

A sleep cycle lasts for 90 minutes. At the end of each cycle, you can naturally wake from sleep. The answer is therefore based on working out when a 90-minute cycle would end. Assuming Spencer is an adult, he will most likely sleep 7–8 hours per night. Therefore, 7.5 hours sleep would have him waking at 6.30 a.m., at the end of five cycles. **A** is incorrect because sleep cycles last for 90 minutes. Therefore, if going to bed at 11 p.m., 6 a.m. would have Spencer waking up part-way through the fifth cycle. **C** is incorrect because sleep cycles last for 90 minutes. Therefore, if going to bed at 11 p.m., 7 a.m. would have Spencer waking up part-way through his sixth cycle. **D** is incorrect because if Spencer slept until 7.30 a.m., he would have had 8.5 hours sleep, and would still be waking up halfway through a cycle.

Question 26

A NWC; high; low

Beta waves are prominent in normal waking consciousness and highlight alertness. They therefore occur very frequently, close together, and are low in height/amplitude. **B** is incorrect because while beta waves occur in normal waking consciousness, they are high in frequency and low in amplitude. **C** is incorrect because beta waves do not frequently occur in an altered state of consciousness (ASC). **D** is incorrect because beta waves do not frequently occur in ASC. Beta waves are high in frequency and low in amplitude.

Question 27

D The proportion of time spent in REM sleep significantly decreases from infancy and then remains steady as we continue ageing.

Newborns and infants can spend up to 50% of their sleep in REM sleep. This changes to about 20% REM sleep as we age. **A** is incorrect, as the amount of time spent sleeping decreases as we age. **B** is incorrect because the proportion of NREM sleep remains relatively stable across the life span, at about 80% of total sleep time. The type of NREM sleep changes across the life span, but not the overall proportion. **C** is incorrect because NREM stages 1 and 2 as a proportion of our sleep time increases as we age.

Question 28

C hallucinations

Hallucinations are an extreme effect of prolonged sleep deprivation and are rare in the first 24 hours. **A** is incorrect because it is likely that India will experience difficulty concentrating, especially on simple tasks. **B** is incorrect because it is likely that India will experience droopy eyelids due to her sleep deprivation. **D** is incorrect because it is likely that India will experience difficulty completing simple tasks (ability with simple tasks is more affected by sleep deprivation than ability with short complex tasks).

Question 29

A alpha and theta waves

A microsleep is a short burst of sleep, and so brainwaves are similar to those experienced when going into NREM sleep. **B** is incorrect because delta waves are associated with deep sleep (NREM stage 3). Microsleeps do not involve entering deep sleep. **C** is incorrect because beta waves are associated with being awake and alert. **D** is incorrect because sleep spindles occur in NREM stage 2, not in a microsleep.

Question 30

B the higher the BAC, the greater the reaction time.

This question is tricky, because you usually interpret graphs by their axes, but this graph has an extra variable. You can see that as each BAC line gets higher, the reaction time gets higher/slower. It also gets worse with age. **A** is incorrect because we cannot conclude a person's state of consciousness from a graph looking at one measurement (reaction time). **C** is incorrect because reaction time is higher/worse as a person ages; this can be seen on all the BAC lines. **D** is incorrect because the graph shows that reaction time gets worse as BAC increases.

Question 31

A age; reaction time

As the researcher is measuring reaction time, this is the dependent variable. Amount of alcohol could be an independent variable, but age is an independent variable too. **B** is incorrect because reaction time is what is being measured, not what is being manipulated, so reaction time is the dependent variable. BAC could be an independent variable. **C** is incorrect because the amount of alcohol consumed could be an independent variable, not a dependent variable. Also, cognitive performance is being measured as reaction time, so it is a dependent variable, not an independent variable. **D** is incorrect because both of these could be considered independent variables, not dependent variables.

Question 32

D estimated the study going for a shorter or longer time than it did.

One of the indicators of an altered state of consciousness is an altered experience of time, either shorter or longer. **A** is incorrect because being in control of emotions would be indicative of normal waking consciousness, not an altered state of consciousness. **B** is incorrect because recalling a stream of thoughts throughout the day would indicate higher levels of awareness and no cognitive distortions, and therefore normal waking consciousness. **C** is incorrect because recalling conversations of passers-by would indicate a high level of awareness and therefore suggest normal waking consciousness, not an altered state of consciousness.

Question 33

D quantitative and qualitative

The first part of the study was quantitative or objective (BAC), but the thoughts were qualitative or subjective. **A** is incorrect because BAC is not subjective, but can be measured objectively as a form of quantitative data. **B** is incorrect because BAC is not a self-report, it is measured objectively. **C** is incorrect because the gathering of data on thoughts and feelings is subjective or qualitative, not quantitative.

Question 34

C the 0.10% BAC condition, cognition affected

It is 0.10% for 24 hours' sleep deprivation and there is a similarity in cognitive effects. **A** is incorrect because while the BAC comparison is right, the effects of sleep deprivation are similar to the effects of BAC on cognition, not on sleep. **B** and **D** are incorrect because 24 hours' sleep deprivation is equivalent to 0.10% BAC, while 0.05% BAC is comparable to 17 hours of sleep deprivation.

Question 35

A their circadian rhythm increases body temperature during the day.

Melatonin is a problem, but it is not released (increased) until night-time. Body temperature is another circadian rhythm that helps signal sleep, and this is the more likely reason shift workers find it hard to sleep during the day. **B** is incorrect because melatonin is not released until night-time; if it were released throughout the day, a shift worker would find it easier to sleep. **C** is incorrect because while noise can make it more difficult to sleep, living somewhere noisy is not universal to shift workers. **D** is incorrect: while shift workers may not feel tired due to lack of melatonin, shift work does affect a person's circadian rhythm, so **A** is the correct response.

Question 36

A low self-efficacy

By process of elimination, low self-efficacy is the correct response. It involves a person's thought processes and beliefs, which are psychological. **B** is incorrect because genetics is a biological risk factor. **C** is incorrect because loss of social network is a social risk factor. **D** is incorrect because resistance to medication is a biological risk factor.

Question 37

C social

Stigma is a social risk factor, as it relates to how others treat or view you. **A** is incorrect because stigma does not involve bodily processes, so it cannot be biological. **B** is incorrect because stigma is based on how others view and label you, whereas psychological risk factors originate within a person and involve their thoughts, cognitive processes, emotions and so on. **D** is incorrect because genetics are a biological risk factor involving a person's genetic makeup.

Question 38

B insufficient sleep

Poor-quality sleep can have negative effects on a person's overall mental wellbeing. **A** is incorrect because support from family is likely to increase a person's resilience and allow them to effectively cope. **C** is incorrect because support from social networks is likely to increase a person's resilience and allow them to effectively cope. **D** is incorrect because an adequate diet is a biological protective factor and can promote resilience.

Question 39

B so that the phobic stimulus can be exposed in increasing intensity

An association is formed between stimuli, not responses, and time is necessary so exposure can start small and intensify. Moving too quickly can have the opposite of the intended effect. **A** is incorrect because while the aim is to associate the phobic stimulus with relaxation, sometimes an association can occur due to one pairing, so this is not the most correct response. **C** is incorrect: systematic desensitisation involves classical conditioning, and therefore does not involve consequences, as this is an operant conditioning principle. **D** is incorrect because systematic desensitisation can be done on people under the age of 18, and it is important to treat a phobia early.

Question 40

D holistic; culture and history

A strength of this framework is its holistic nature and the emphasis it places on culture and history. **A** is incorrect because this framework is holistic in nature, not specific. **B** is incorrect because this framework is holistic in nature, not specific. **C** is incorrect because while the framework is holistic, it focuses on the importance of culture and history, not brain and body.

Question 41

A acculturative stress; daily pressures

This one is tricky because of the need for interpretation. Tuan managed his money differently in his birth country (acculturative), while small incidents at work that we all encounter could be considered daily pressures. **B** is incorrect because chronic stress is prolonged and severe, whereas the stress Tuan is experiencing with his housemate is likely due to cultural differences and is therefore acculturative stress. At work, while Tuan may be experiencing distress, this is likely occurring due to daily pressures. **C** is incorrect because these responses are around the wrong way. **D** is incorrect because major stress is not being experienced at work; major stress involves extremely traumatic experiences.

Question 42

A Tuan is fearful when the outcome is not predictable.

Fear due to lack of control can be a symptom of anxiety. **B** is incorrect, because Tuan's stress is inconsistent. **C** may occur later for Tuan, but this statement is not an accurate reflection of the scenario. **D** is incorrect, because contributing factors don't lead to diagnosis.

Question 43

B inhibit the stress response.

Benzodiazepines stimulate GABA production, which in turn inhibits the stress response and calms the body. **A** is incorrect because benzodiazepines do not block the stress response entirely, but have a calming effect on the central nervous system. **C** is incorrect because benzodiazepines have a calming effect, mimicking GABA production and making it less likely for the stress response to occur. **D** is incorrect because benzodiazepines work by stimulating and mimicking GABA production.

Question 44

C an individual's right to shape their own life so they live according to their values

Self-determination is important because Aboriginal and Torres Strait Islander peoples determine their own sense of wellbeing. **A** and **D** are incorrect because they are not theoretical knowledge. **B** is incorrect: this is the definition of cultural continuity.

Question 45

C Aboriginal and Torres Strait Islander cultures have a stronger emphasis on connection to Country than Western cultures.

Both Western and Aboriginal and Torres Strait Islander cultures explore mind and body, but the exploration of healing in Aboriginal and Torres Strait Islander cultures places a stronger emphasis on Country. **A** is incorrect because both Western and Aboriginal and Torres Strait Islander cultures explore mind and body. **B** is incorrect because Aboriginal and Torres Strait Islander cultures place a greater significance on connection to Country than Western cultures.

Question 46

A qualitative; interviews; small

Although the type of data, data collection method and sample size in **A** and **C** were congruent, they were best matched to a qualitative case study methodology, so the answer is **A**. **B** is incorrect because quantitative data is objective and involves numerical values rather than descriptions from interviews. **C** is incorrect because interviews would provide richer descriptive data than a questionnaire. Interviews also tend to be conducted on smaller samples. **D** is incorrect because quantitative data is objective and involves numerical values; it will not give rich descriptions.

Question 47

D change score calculated as the difference between pre-simulation and during-simulation stress level scores.

Here, we need to infer that stress level was operationalised as a single change score derived from averaging the two separate change scores that could be calculated for each of the electromyograph (EMG) and EMG arousal scores. The change score was the best response, because Ravi's aim was to find out whether the coping strategy used would 'affect their baseline levels of stress'. The change score operationalises the difference between baseline and coping strategy stress levels. **A** is incorrect because the dependent variable is what is being measured – in this case, the change in scores, not the level of coping flexibility. **B** is incorrect because students had to infer that stress level was operationalised as a single change score (not just the score) derived from averaging the two separate change scores that could be calculated for each of the electromyograph (EMG) and EMG arousal scores. **C** is incorrect because we need to infer that it was the change in scores pre- and post-simulation, not just the level of arousal.

Question 48

A using the same participants in both conditions, as there may be practice effects

This result indicates a need for better understanding and application of the precise definition of the term 'confounding variable'. A confounding variable is a variable that has the potential to directly and systematically affect the measurement of the dependent variable. **A** was expressed in generic terms, but was the only option that relates to a true confounding factor, rather than to the effect of a bias, a problem with generalisability, or artificial restriction of available coping strategy. **B** is incorrect because not being able to generalise the results is a limitation of a study, rather than a confounding variable. **C** is incorrect because this may bias the participants, but it may not necessarily systematically influence the independent variable. **D** is incorrect because the experiment was based on participants using a particular strategy in each condition.

Question 49

A detect changes in levels of arousal.

The dependent variable is dependent on measuring the change that occurred. **B** is incorrect because the dependent variable is what is being measured, and therefore it is looking at changes in level of arousal. **C** is incorrect because reliability involves the study needing to be replicated. **D** is incorrect because these participants' results were not eliminated.

Question 50

B +6.2; +1.5

We are looking for a higher score in avoidance than approach, and for an overall score that is an increase not a decrease. **B** is the only option. **A** and **C** are incorrect because for support to be gained for Ravi's hypothesis, there must be a higher score for the avoidance strategy. **D** is incorrect because although there is a higher avoidance score, the approach is a decrease so he is not comparing two increases.

Short-answer sample responses

Question 1

a A neurotransmitter is a chemical substance secreted by a neuron to allow neural communication to occur, or a particular function of the body. (1 mark)

Glutamate is one such neurotransmitter. Too little glutamate may make it difficult to learn or concentrate. (1 mark)

Too much glutamate can lead to overstimulation, and then to insomnia or restlessness, due to a feeling of being 'wired'. (1 mark)

b Dopamine can act as a motivator, driving you to get something done that you will feel good about, such as finishing your homework. (1 mark)

Dopamine can be harmful when it supports addictive behaviours, such as the urge to gamble and the inability to stop, despite losses. (1 mark)

Question 2 ©VCAA 2020 EXAM REPORT SB Q2

a That participants/volunteers who are provided with information about the positive effects of stress (responses) will have a lower level of stress during public speaking (smaller difference in heart rate after giving a speech from baseline) than participants/volunteers who are not provided with information about the positive effects of stress (responses). (3 marks)

b The researchers measure heart rate in this study because an increase in heart rate is an objective or quantitative measure of the sympathetic nervous system's response to stress. (3 marks)

c Group 1 – Experimental group

- Secondary appraisal: The experimental group would be more likely to appraise themselves as having the resources they need to cope with the (challenge of) the speech.

- Justification: They received information that taught them that experiencing a stress response means their body is ready to support them for the challenge of public speaking. Evidence that this occurred can be seen in the smaller change in heart rate from baseline for the experimental group. (3 marks)

Group 2 – Control group

- Secondary appraisal: The control group were less likely to appraise their resources as being sufficient to cope with the stressor (threat) of giving the speech because they did not receive information about the positive effects of stress.

- Justification: Evidence that this occurred can be seen in the larger change in heart rate from baseline to immediately after the speech for the control group. (3 marks)

d Having the majority of athletes randomly allocated into the experimental group produces a confounding variable. The effect of the independent variable on the dependent variable cannot be unambiguously attributed to the intervention because athletes' heart rates are likely to be slower under stress than those of non-athletes. Therefore, having the majority of athletes in the experimental group confounds the explanation of the observed difference between the groups in post-speech heart rate because athleticism provides an alternative cause of the observed effect. (3 marks)

Question 3

a The gut–brain axis tells us that gut health can affect mental wellbeing and vice versa. (1 mark) If Penny has some bad bacteria in her gut, it may influence her feelings of stress. (1 mark)

b • Approach strategies help deal with the problem cognitively, or by dealing directly with the stressor. (1 mark)

  • They are effective at relieving stress. (1 mark)

  • Avoidance strategies involve ignoring the problem and doing nothing about it. (1 mark)

  • Although it may feel momentarily good, it is not an effective technique for managing stress. (1 mark)

Question 4 ©VCAA 2020 EXAM REPORT SB Q4

a The advertisement uses classical conditioning principles (1 mark) because the drink acts as a neutral stimulus that is repeatedly paired with unconditioned stimuli (images of people having fun) that promote a positive emotional response. (1 mark) With repetition, the drink becomes a conditioned stimulus that can produce positive feelings on its own. (1 mark)

b Process 1: Attention (1 mark)

Justification: The viewer is more likely to actively focus on the drink and encode information about it if it is associated with someone famous. (1 mark)

Process 2: Retention (1 mark)

Justification: Viewers are more likely to remember the product and the pleasant feelings associated with it because it is linked in memory with someone they admire. (1 mark)

Question 5

a Iconic memory stores visual information, whereas echoic memory stores auditory information. (1 mark)

b They are both sensory memory stores (or they both have unlimited capacity). (1 mark)

Question 6

One mnemonic from written cultures is an acronym, where the first letter of every word to be remembered creates a word. (1 mark) An advantage of this is that it can be quickly constructed to aid memory. (1 mark)

One mnemonic from oral cultures is songlines: a narrative carried through songs in communities. (1 mark) It is advantageous due to the detail that can be included. (1 mark)

Question 7

Any three of the following

• REM represents about 20% of total sleep, whereas NREM represents 80% of total sleep.

• You experience beta-like waves in REM, whereas in NREM you experience alpha, theta and delta waves.

• Most dreaming occurs in REM; it is rare and less sensical in NREM.

• Your body is paralysed in REM, whereas you can move in NREM.

• You get more REM as the night progresses; you get less NREM as the night progresses.

Question 8

a An EEG detects, amplifies and records electrical activity of the brain. (1 mark)

b An EOG detects, amplifies and records electrical activity of the muscles around the eyes. (1 mark)

c Near brain (EEG); along the sides of the eyes (EOG); on the lower sides of the face (EMG) (1 mark)

Question 9

a Any two of the following:

- Gemma might have less inhibition, leading her to do things that are reckless.
- Gemma may have poorer memory, so she may forget some road rules.
- Gemma may have compromised cognitive skills, so she may make poor decisions.

b The negative effects associated with being sleep deprived for 24 hours are the same as having a BAC of 0.10. (1 mark)

Question 10

Educators are worried about students not getting enough sleep, as this can have a negative impact on their wellbeing and can be a contributing factor to the development of mental illness. (1 mark) Electronic screens emit blue light, which suppresses the release of melatonin. (1 mark) Without the release of melatonin, people don't feel sleepy when going to bed. (1 mark) This is why screens should be avoided 2–3 hours before bed. (1 mark)

Other advice to help the sleep–wake cycle could include having a bedtime routine, going to bed at the same time each night, utilising natural light cues, and limiting food and caffeine late at night. (2 marks)

Question 11 ©VCAA 2018 EXAM REPORT SB Q4 (ADAPTED)

a 'People who undertake regular exercise will show a reduction in the normal age-related decline of brain structures related to memory and learning compared to those who do not undertake regular exercise.' (3 marks)

b Students needed to recognise that the study sample included only healthy older adults, and so the results could not be extended to those with Alzheimer's disease. (1 mark)

c Students needed to demonstrate knowledge of the following in their response (3 marks):

- (The earliest stages of) Alzheimer's disease is associated with a reduction/impairment of hippocampal volume/function (due to neurofibrillary tangles and amyloid plaques) affecting learning and memory (especially short-term and episodic memory, anterograde amnesia).
- The effects of moderate intensity aerobic exercise seen in the study suggest that it can protect against decline in hippocampal volume.
- Therefore, researchers may be interested in how regular moderate intensity exercise might act as a protective factor against the development of Alzheimer's disease.

d Students needed to demonstrate knowledge of the following three elements of consent when conducting research with people who have a cognitive impairment (3 marks):

- the role of informed consent is to ensure that the participant understands and willingly agrees to the procedures involved in the experiment, including understanding the potential harms/risks (withdrawal rights, how their data may be used)
- those who have Alzheimer's disease may not have full cognitive capacity to give consent (depending on the stage of the disease)
- the appropriate extra protocols required to ensure informed consent from persons with a cognitive impairment, especially the requirement to obtain consent from the person's (legal) guardian or another person/organisation authorised by law. (This may need to be revisited if the study continues over a period of time and the person's cognitive impairment worsens.)

Question 12

Biological: GABA dysfunction – having too little to inhibit the stress response and long-term potential of a critical incident that triggers neural pathways (2 marks)

Psychological: Acquiring a phobia through classical conditioning and then continuing it due to operant conditioning and cognitive biases, such as memory bias (only seeing the bad things) and catastrophic thinking (escalating to negative outcomes) (2 marks)

Social: Environmental triggers that strengthen the phobia, and not seeking treatment due to the stigma that occurs (2 marks)